Life-Sustaining Organizations

A Design Guide

Michael J. Sales, Ed.D. and Anika Ellison Savage

Art of the Future

Copyright © 2010 by Art of the Future™, Anika Savage and Michael Sales

All rights reserved. No part of this book may be reproduced or utilized by any means, electronic or mechanical, including photocopying, recording, or by any information storage and retrieval system nor may it be transmitted, nor translated into machine language without the written permission of the copyright holder. Your support of the author's rights is appreciated.

The authors may be contacted at Art of the Future, info@artofthefuture.com.
    Sales, Michael J. Ed.D. and Savage, Anika Ellison

Life-Sustaining Organizations: A Design Guide
    / Michael J. Sales Ed.D. and Anika Ellison Savage

    Includes Influences and Resources and Index.

    1. Future of work 2. Structural dynamics 3. Systems thinking 4. Scenario analysis
    5. Strategic planning 6. Living systems 7. Workplace strategy I. Title

Library of Congress Cataloging Number: 2010909269

EAN-13: 9781453633083
ISBN-10: 1453633081

For information regarding special discounts for bulk purchases, contact Art of the Future.
                    info@artofthefuture.com

This book is dedicated to
Leila Sales and Cam Schriefer
with lots of love, respect and gratitude.
You make us excited about the future!

Acknowledgements

We appreciate the support and encouragement of Nola's Salon with thanks to Grady McGonagill, Marilyn Darling, Anne Starr, Peter Pryun, Eleanor Seamans and, especially, our associate Rick Karash.

We recognize the contribution of thought leaders we have known and learned from over the years. These include Peter Senge, Barry Oshry, Peter Schwartz, Michael Marien, David Cooperrider, Michele Bowman, Myron Sharaf, Jim Dator, Margaret Wheatley, David Millen, Stanley Saltzman, Christopher Alexander, Jay Vogt, Chris Mackin, Bob Rosenthal, Liam Fahey, Arie de Geus, Otto Scharmer, Lee Bolman, Joan Gallos, Michel Crozier, Bill Joiner, Barry Dym and Chris Argyris. Their perspectives and theory building skills have profoundly influenced the evolution of our thinking. Our special thanks goes to Doug Dowd, an early influence in the art of comprehensive systems thinking.

We have had the great fortune of working with anticipatory leaders such as Angelo Lamola, Sam Felton, Ron Newbower, Jeff Cooper, Rita Cleary, Sherry Immediato, Sharon Wulf, Bryan Berthold, Ben Chirgwin and Harvey Greenberg.

We have collaborated with many of the leading thinkers and practitioners in the development of workplace theory and practice including Marty Anderson, Jeff Austin, Charlie Grantham, Jim Ware, Fritz Steele, Chris Hood, Andrew Laing, Peter Miscovich, Keith Perske, William Porter, Cynthia Frogatt, Michael Joroff and Joel Ratekin.

We value the contribution of wonderful colleagues to our intellectual, professional and spiritual development including Jonathan Milton, Mark Horowitz, Anne Litwin, Joe Meier Mary Lou Michael, Phil Novick, Penny Ford-Carleton, Jyoti Ganesh, Teressa Griffith, Jane Critchlow, Kevin Purcell, Napier Collyns, Bill Wilson, Karen Oshry, Steve Stulck, Lisa Hirsh and many others who we could [and probably should] name.

We are very thankful to friends and supporters who play a role in all the many positive features of our lives, including John Hanson, Paula Beall, Dorothy Ahlgren, Jack Schaeffer, Harry Geisberg, Esther Sales, Bob Rosenthal, Katherine Deutsch Tatlock, Mary Lee Kennedy and Michael Charney. We particularly recognize the contribution of John Ahlgren, leader, teacher and poet, to our growth and well-being.

We remember Stephen Sales whose impact on his younger brother was multifaceted, of course, but particularly for his experimental social science design skills. He looked where no one else thought to go. We need more of that kind of curiosity and foresight. We appreciate the strength, support and friendship of Heidi Ellison, a shining light in our lives. And we recognize the other members of our families whose love and belief in us is reciprocated every day.

...and we acknowledge each other for our enduring patience, insightful conversations and continual understanding throughout this creative process. Through our collaboration, our ideas have flourished and combined into a wholeness that enabled us to create this guide.

Preface

> People only come alive to the extent that the buildings, organizations and towns they live in are alive.
>
> *Christopher Alexander, 1979*

Our Organizational Stories

Great work experiences can be extraordinarily fulfilling, deeply meaningful and truly life transforming. Working within organizations, we have had opportunities to contribute our knowledge, talents and skills to affect a larger context than we could have done as individuals. We have been appreciated and rewarded for our contributions, even as the process has enlivened us. *And,* too often, we have realized that much more was possible at the personal, organizational, economic, societal and planetary levels than our organizations had accomplished. Our own experiences with organizations large and small have made us poignantly aware of the enormity of the human potential that is unfulfilled in our working lives.

We are compelled to write this Guide to describe a path to healthier, more productive alternatives to the way things are currently being done at work. Our experiences have inspired us to create an approach that organizations can use to reshape their thinking to better meet the demands of the future as we move from the Industrial Era into the new, Organic Era that is emerging.

- **Anika's Story: a Tale of Two Organizations**

    My very different experiences as a professional employee of two organizations caused me to consider what made for such stark contrasts, what lessons I could draw from them and how I could help make worklife a consistently better experience for both employees and employers. I have since dedicated my career to finding the answers to these questions, and I was motivated to create this Design Guide to share what I've learned so far.

    The Right Place?!
    Early in my career, I worked for a series of small firms, each no larger than a dozen people. During an economic downturn, I sent my resume to a high-flying technology company that was looking for someone with my skills. For several years, the entire front page of the "Help Wanted" section of the Sunday *Boston Globe* routinely featured this firm. With 26,000 employees, the company seemed to employ everyone in the surrounding communities. I was selected from a pool of 300 applicants for a single position. The company had a reputation for innovative products, for its networked management structure and for allowing its employees a high degree of autonomy. "If you propose it, you own it" was a popular mantra. Revenues and profits grew by double digits year after year. The business and popular press, even Hollywood, was impressed by the company's phenomenal performance and game-changing products. So, I thought I'd give it a try – for a month, maybe two.

Preface

The facility where I went to work was a two-story structure, made of concrete with tiny windows enclosing 400,000 square feet of space. The interior walls and cubicle panels were beige. The "private" offices were seven feet high, open at the top, with the walls stopping three feet below the suspended ceiling. Most of us had five-foot-high panels with no doors, no acoustic privacy and no visual connection to coworkers. My initial sense was that the people working there were automatons, otherworldly beings quietly shuffling about. They seemed a perfect fit for this environment. Or, was I the one who had just landed from Mars?

On my third day, I was finding my way to the cafeteria at the other end of the building by carefully noting the "pole numbers" on the columns [B1, B2, C2, etc.] when one of these robotic types entered into my personal space and seemed to be human. He introduced himself as "my HR representative" and wanted to know if I had any problems. "Not that I know of" was my short and shocked reply. He was visibly disappointed. It felt a bit like a hit-and-run psychological checkup. Later that day, badge number 3 stopped by to introduce himself. Since I was number 51749, I knew he must be "someone." He turned out to be the founder's brother and the executive in charge of the project I was working on.

Over time, I came to learn about personnel, matrix management and many other useful things – including pension plans [remember those?]. In this plan, you became "vested" in the fifth year of employment and reached 100% after 10 years. My time horizons gradually stretched from months to five years, then another year, then 10, and still I stayed on. I was one of 125,000 employees when I hit a psychological wall after 18.4 years [but who's counting?] and knew I had to leave. My positions had changed during that time with growing levels of responsibility and my compensation had kept pace. By most measures, I should have been a dedicated, engaged employee. So what happened?

The truth is that I rarely experienced a comfortable "fit" between myself and the organization. I am a highly creative person, but that creativity showed up less and less in my work – it wasn't really needed or desired. I conformed in many ways, both obvious and subtle. Family life was pigeonholed into nights and weekends. If you took two consecutive weeks of vacation, you were admitting that you weren't really needed; having unused vacation time at the end of the year was a sign of commitment. Maternal bonding with my newborn had to be completed in six weeks. So I could get back on the job in the requisite amount of time, I had to leave my baby to be cared for by strangers in an institutional setting. On my way to work, only twenty minutes from home, I passed shops that opened later than my morning commute and closed earlier than my return in the evening. I lived in a beautiful New England town, but I felt I was simply viewing it without any opportunity to become a part of the community. Accepting an invitation to join the local garden club was ridiculously out of the question – I didn't have time to smell flowers let alone grow them. I had been contorting myself for the sake of my job for so long, in so many ways, that I had forgotten who I was or whom I might have become without it. The organization was controlling and sapping my life.

Intermission
I felt as though I was jumping off a high, craggy cliff into a frigid, tumultuous ocean when I resigned. I threw myself into the arms of the universe. Fortunately, the universe caught me in midair and set me down very nicely. I landed on my feet as an independent consultant on a two-year contract with one of America's oldest corporate giants. When they offered me a position as an employee, I didn't even consider it. I was

Preface

INDEPENDENT and earning a great income to boot. While my clients sweated over changes in senior management, reorganizations and spin-offs, I blissfully focused on our project work. I thought I had died and gone to heaven.

The Wrong Place?!
Several years later, economics compelled me again into full-time employment. I accepted a job in the financial sector, to start in one month's time. The intervening month was to be "my" time – to do whatever I wanted. Instead, I spent the entire time in panic mode, often sobbing. I just knew I was selling my soul. My new company was an institution with poor customer relations, a tough-minded CEO and a history of devouring one bank after another as it grew in size and market share. I braced for the worst – I would put my true self back into hibernation to endure this necessity, grit my teeth and bear it. To my great surprise, I loved it! What happened?

The urban location made driving to work difficult, time consuming and expensive. However, it also meant that it was a hub for public transportation. Because I had a choice of commuter train, bus or subway, I had flexibility in planning events at either end of the day. And the ride provided time to read, work, mediate or just think. At the office, people greeted one another in the corridors. My office, although still a cubicle, was in a tower with large windows and panoramic views of a major, bustling metropolis. The environment incorporated color and personal effects. Our team sat near people from other departments on whom we depended, reducing the "us" and "them" mentality. People at all levels expressed a high degree of consideration and appreciation for others. Executives recognized, respected, accepted and rewarded our work. Hours and location were flexible enough to accommodate personal needs. We almost never took sick days; rather, we called in "working at home" and were, indeed, on the job during the day. Severe weather conditions were also accommodated in this way.

I had an excellent working relationship with my manager. We had weekly one-on-one meetings. We used this dedicated time to discuss administrative matters, personnel issues and project status – whatever needed attention. The effect was so positive that I initiated the same process with my direct reports. As a result, healthy working relationships developed, and any unpleasant surprises on projects or at review time were eliminated. The department also encouraged quarterly group meetings and held semiannual departmental meetings that included all employees. We shared in the good times and bad. When the "tech bubble" burst, employees readily accepted a notable change in the level of rewards while recognition practices remained unchanged.

So I wonder, what made my first organizational experience so painful and the second one so positive. As I look over what I've written here, I have to conclude it was the people: their interpersonal and professional skills, their commitment to their work, their respect for others and their sense of contribution to the organization and the wider world. These characteristics in ourselves and our colleagues change work from drudgery to a fulfilling experience.

- **Michael's Story: Too Many Unhappy People**
Having come from a family of entrepreneurs, I could have joined my Dad's business but decided to go in my own direction. I majored in finance at Wharton and worked as a securities analyst for a Wall Street advisory firm while at the University of Chicago School of Business. I was hired as a junior systems analyst by an insurance company in 1968. The firm invested premiums in financial instruments, and my investment

Preface

background was an advantage. For example, based on my analysis of an equity, my supervisor made $10,000 in two weeks by trading the stock. His assessment of my performance was "exceeds requirements." Therefore, it was really shocking when the same guy unexpectedly terminated me soon afterward. The reason given: dandruff!

I felt lousy being vulnerable to such capriciousness. That experience, at age 23, made self-employment attractive. That was my last job as a salaried employee, but I didn't stop working:

- I joined a community newspaper housed in an old office building in Santa Cruz, California. The town and surrounding area were beautiful, but our workspace was old, cramped and funky. It didn't matter. Our dedicated team did wonderful work there infused with great spirit. We made a lasting contribution to the culture of the community.
- As a doctoral student, Harvard was my work environment. I loved the physical space and felt particularly connected to certain settings, like the reading rooms at the Widener and Baker Libraries, the Radcliffe Quadrangle, some small study rooms and the museums. They contributed to my feeling of being in a special place. The distinctive, historic context filled me with a sense of responsibility. Everyone knew that many scholars had done their best work at Harvard; this knowledge inspired me to do the same.
- For 25 years, while also working as a management consultant, I was on the staff of the Power Lab, a highly creative experiential workshop. This week-long residential program is held at an 1880s vintage resort on Cape Cod. The gorgeous grounds and rustic accommodations create a backdrop that is a central ingredient of the transformational process experienced by many participants.

The newspaper, Harvard and the Power Lab – these were joyous experiences! I have worked as an independent organizational development consultant for more than 30 years in healthcare, manufacturing, technology, insurance, financial services, investment banking, small business, secondary and higher education, professional services, transportation, government, consulting and other settings. I have worked directly with thousands of people and had numerous great teamwork experiences with both clients and colleagues. A real feeling of partnership made many of these work experiences extraordinarily vital. I flourished in these situations and learned much about the amazing possibilities of working life.

Among my clients, I have certainly found individuals and teams reporting satisfaction with their jobs and work experiences. Unfortunately, these were the exception rather than the rule. Why? Here are a few impressions.

- The physical environments in the vast majority of client settings have been alienating. The facilities seldom possess a unifying aesthetic; if one exists, it has been imposed by professional designers and doesn't reflect anything unique about the team and its members. Music is limited to occasional Muzak in public spaces or the sporadic radio playing rock music too loudly. I once coached a corporate director who said that her office made her feel like a prisoner.
- People frequently feel that only a small fraction of their real selves is welcome at work. Conformity to limiting rules about self-expression takes precedence over the freedom to make innovative contributions by taking risks. This sort of unspoken bending of the self to organizational norms is so commonplace that most people don't even recognize it.

Preface

- Many people feel disconnected because organizations communicate powerful messages regarding relative status. The majority of workspaces exist in bland, cavernous enclaves of cubicles with little or no exposure to natural light. Those with "status" demonstrate their rank by occupying private offices along the exterior perimeter; the highest-ranking executives in the corners. I have eaten in many crowded, noisy cafeterias and in drab vending-machine and microwave lunchrooms, sometimes informally segregated by organizational roles [white, blue and pink collar]. In many of these settings, posters touting the benefits of teamwork or graying placards with the organization's vision, mission and/or values often pass for graphics. The message these spaces convey is, "Hurry up, get back to work." On the other end of the spectrum, I once had dinner with a group of investment bankers inside a vault, which was the ultra-private room at an already exclusive restaurant. Everything about the experience communicated how special they felt they were. These stratified environments create barriers that prevent people from getting to know each other across levels and functional boundaries. Working relationships develop in spite of, rather than because of, the facilities intended to support such activities.

- Recently flying over the Great Lakes, I asked the attendant, "Is that Lake Erie out there?" She responded, "Couldn't tell you. I never look out the window." Few people pay attention to anything lying outside their organizational windows. Even though enormous crises and thrilling developments occur in virtually every aspect of life, most organizations give little thought to the forces shaping their context. I was once present at a conversation on the status of an organization's lobbying efforts, but I have never been at a thoughtful, multidimensional discussion of any major political issue, its implications or the potential range of organizational responses to the topic. For example, I have heard many people complain about traffic hassles to and from work, but I've never heard any discussion of an organization's stance on transportation policy. I have noticed people complaining about crime and fretting over immigration policy, but I've never attended a meeting that probed the roots of either of these problems or that included a conscious reflection on what the organization's stance should be. One of the most talented executives I have worked with confided that he had not read a newspaper cover-to-cover during the entirety of his 35-year career. The message seems to be: "Let someone else worry about what's out there; we've got work to do." The fact that their organizations influence – and have the power to change – many external conditions doesn't seem occur to most people.

- Ecology is an afterthought in most of the organizations I've worked with. Companies seldom express explicit or widely affirmed values regarding recycling, energy and water conservation or managing their carbon footprint beyond compliance requirements. For example, for 18 months, I had a job at a corporate residential training facility. Using a rough calculation, I estimated that the lights left on full time in the meeting, conference and training rooms, dining facility and corridors consumed enough energy to illuminate over 500 houses daily! In another case, an energetic employee intent on improving her organization's recycling efforts initiated a program in a large office facility. When she left, the program faltered and largely disappeared. All too often, organizations regard the natural environment as an endless resource and a dumping ground.

Preface

- As a coach, I've had the opportunity to get to know several hundred people quite well. Nearly all of them have a spiritual life of some kind. Most are not religious in a traditional sense but they are questioning, searching and believing in something beyond materialism and consumption. None of the organizations I've worked with, aside from their charitable and civic activities, provide any way for people to manifest this aspect of themselves at a personal or group level, such as encouraging even a few moments of silence.

- Executives in our client organizations are definitely moving away from hiring only those who look like themselves. Firms are becoming more diverse for a variety of reasons – regulatory, practicality, strategic positioning and, most important, because they see the benefits. *And*, there seems to be plenty of talent on the periphery of organizations that hasn't yet been tapped [all those bright energetic consultants that organizations use but never hire because they are "too independent"]. It seems to me that the organizational mold that determines a person's fit could be stretched to embrace a broader range of talent, among both existing and prospective employees.

There are important differences in the stories of our working lives. However, we have arrived at a very similar perspective: the world of work is a grayer, lonelier and sadder place than it could be, and that's not good enough...not now and not going forward. We can and we must do better to ignite job satisfaction, to enhance productivity, to satisfy customers, to sustain the environmental and to improve the bottom line.

Our Purpose in Writing

We believe that humanity is on the brink of extreme possibilities. One extreme is full of opportunity: a dramatic evolution of our species and marvelous advances that we can hardly imagine. The other is a calamity for civilization: continued assault on the natural environment and a decline in human existence on the planet. We are committed to playing some small part in the creation of a life-positive future. We are writing to encourage organizations to consider the vitality of the work environments they've created and to show how to infuse them with fresh energy so that they are life-sustaining rather than life draining. We have experienced over and over again that organizations that support the essence of their employees are much more productive. This Guide presents a rigorous approach that organizations can use to consciously become living systems, sustaining the people who bring the organization to life, now and in the future.

We believe that work can be an ongoing peak experience, an unfolding of creative expression. Outside of work, people are dancing, playing music, volunteering, shaping communities, schools and churches, painting, drawing, singing, expressing ourselves and developing our potential in all sorts of ways. We'd love to see organizations embrace the enthusiasm for life that is so human, so that our work systems become as resplendent as our natural systems. Writing this Guide is a life-sustaining work experience for us – a dream come true.

How to Recognize a Life-Sustaining Organization

Consider an organization you care about – as a leader, an employee, an investor, a customer, a partner or simply as an observer. Is it life-sustaining? For the organization, rate the following statements as 1. <u>Very True</u>, 2. <u>Somewhat True</u>, or 3. <u>Not True.</u>

1 2 3

- People in this organization are cheerful and energetic; they have an easy rapport and are considerate of each other; they enjoy working together.
- Staff members discuss things and can disagree without being nasty; they alternate between collaborating and working alone in an easy flow.
- A wide variety of people work for this organization. You can tell that there is a lot of creativity. People share and build on each others' ideas.
- The organization innovates new products to new ways of doing things.
- Not only are the people who work here proud of what they do but their families are also pleased to be connected to this organization.
- Everyone who works here can tell me about the organization's mission; they are excited about it. They know the strategies the organization is pursuing and why.
- Employees know how their role supports the mission and strategies of the organization and why their work is important.
- Everyone seems to know what is needed without having to be told in minute detail.
- The organization doesn't have a lot of rules; there is a core set of norms and standards of behavior that everyone buys into.
- It is easy to see that this organization really understands its connection to the natural ecology. The people who work here are aware of their impact on the environment.
- Everywhere I look there Is something intriguing that I want to know more about.
- It feels good to be associated with this organization. I am confident that I am being treated fairly and receiving good value for my time, money, and efforts.
- The look and feel of this organization is pleasing; everything fits together in an aesthetic whole; a lot of attention is paid to every detail of the experience – from the physical space, to products, services and packaging.
- This organization really delivers! It has a great reputation in its field; I have confidence in their products and services.

Now connect the dots that you have selected. What impression does this give you of the organization? If you assess seven or more of these statements <u>Very True</u>, the organization is pretty life-sustaining. If you find four or more statements <u>Not True</u>, the organization probably has some work to do to become life-sustaining.

Table of Contents

Introduction 1
- Organizations as Living Systems — 2
- Approaches to Change — 4
- When to Change — 5
- Life-Sustaining Organizations — 6
- Growing a Life-Sustaining Organization — 15
- How to Use This Design Guide — 17

Part 1: Becoming Life-Sustaining

Convening The Team 19
- Role of Leaders: Integrator — 20
- Preparing for the Life-Sustaining Initiative — 25
- Convening — 31

Exploring Facts 35
- Role of Leaders: Futurist — 36
- Scanning Events — 39
- Recognizing Patterns — 46
- Fathoming Structure — 50
- The Heart of Structural Dynamics: The Scenario Game Board — 54
- Exploring Deeper — 69
- Executive Briefing: Facts — 71

Discovering Options 73
- Role of Leaders: Strategist — 74
- Analyzing Future Possibilities — 77
- Articulating Scenarios — 80
- Living in the Future — 82
- Developing and Testing Strategies — 86
- Executive Briefing: Options — 90

Embodying Action 91
- Role of Leaders: Integrator — 92
- Embedding the Strategy — 95
- From Embedding to Embodying — 98
- Executive Briefing: Action — 106

Sustaining Results 107
- Anticipatory Leadership — 108
- Why Monitor — 112
- What to Monitor — 114
- How to Monitor — 115
- Sustaining Beyond the Stable State — 120

Table of Contents

Part 2: A Case Study

Introducing GoGo Global Transport 124
Convening the Team 128
Exploring Facts 136
Discovering Options 152
Embodying Actions 164
Sustaining Results 170

Final Word   179
- Taking a Deep Dive   180
- The Evolution of Work   181
- Making the Mental Shift   184
- Future Mind   185

Structural Dynamics Process   187

Influences and Resources   191

Image Credits   207

Index   209

Introduction

- **Organizations as Living Systems**
- **Approaches to Change**
  - Fix it
  - Let it collapse
  - Try something new
  - Leap into a new reality
- **When to Change**
- **Life-Sustaining Organizations**
  - Creative People
  - Whole Systems Thinking
  - Design Integrity
  - Elegant Solutions
  - Results Orientation
  - Your Experience
- **Growing a Life-Sustaining Organization**
  - Structural Dynamics
  - Anticipatory Leadership
- **How to Use This Design Guide**

Sunday	Monday	Tuesday	Wednesday	Thursday	Friday	Saturday
get ready for the week	most fatal heart attacks today	count days to retire-ment	hump day	plan the weekend	TGIF	weekend warrior

Introduction

🌐 Organizations as Living Systems

Many people in the workforce of global corporations, small businesses, not-for-profit institutions and government agencies are trying their best, working long hours, living in a continual state of stress, sacrificing their health, fitting home and family life into fragments of time, living in fear of losing their jobs *and* their personal relationships, compensating for colleagues who have lost their jobs and bewildered by who or what is to blame for the poor results their organizations are experiencing. Many leaders do not regard the well-being of the people who give life to the organization as a strategic issue. We believe leaders have no higher priority than doing whatever it takes to support and nurture the full engagement of people in the workplace. They are the heart and soul of the organization. When they thrive, the organization thrives.

> The fundamental problem with most businesses is that they're governed by mediocre ideas. Maximizing the return on invested capital is an example of a mediocre idea. It doesn't uplift people. It doesn't give them something they can tell their children about.
>
> — *Bill O'Brien, Former CEO Hanover Insurance*

This Guide is not about creating happy organizations. Rather, it is a guide to becoming an organization that allows the best in each individual to flourish. When this happens, courage replaces fear, strengths triumph over weaknesses and innovation thrives. New and rediscovered practices closely align with the challenges we are facing in this century. Human, economic and ecological well-being are understood to be inextricably linked – all part of the same fabric.

A compelling vision of the global future must include a harmonious relationship between organizations and sustainable societies. We already have the knowledge, the technology and the skills needed to transform our world into a future that we would want for ourselves and our descendants. Organizations have the power to pull these elements together to realize that future vision. Corporations and other entities are beginning to see themselves as responsible agents for the benefit of global health and longevity. Such organizations are life-sustaining. People love to join them and hate to leave them because they are passionately alive.

> Let us choose to unite the strengths of markets with the power of universal ideals. Let us choose to reconcile the creative forces of creative entrepreneurship with the needs of the disadvantaged and the requirements of future generations.
>
> — *Kofi Anan Former Secretary-General The United Nations*

A fundamental shift in consciousness is occurring: we are becoming aware that our organizations are dynamic, complex organisms rather than thinking of them as machines. Machines are built up piece by piece; the parts are interchangeable. They are designed to perform a particular task or set of functions. They do the same thing over and over. Organizations operating as machines have fixed mental models, allowing them to make decisions quickly. There is little new learning.

Living systems, on the other hand, are constantly growing, changing, self-correcting and evolving. They are dynamic. As human beings, we know that our parts never existed separately from our bodies; we were never assembled. Organizations are starting to recognize themselves to be living organisms, continuously responding to changing conditions and capable of making major leaps in understanding and action. The mental

Introduction

models of these organizations shift in understanding in response to changes in their environment. Living systems have the capacity to adapt and renew themselves.

> There is nothing in a caterpillar that tells you that it is going to be a butterfly.
>
> *Buckminster Fuller*
> *1895 – 1983*

> The caterpillar and butterfly are two temporarily stabilized structures in the "coherent evolution of one and the same organism.
>
> *Jantsch 1980*

Like the caterpillar/butterfly, living systems may undergo phase transitions but their being remains stable. Living systems maintain their integrity over time and across phase transitions, but their precise shape or boundaries will morph. They transform themselves in ways that maintain their integrity without requiring conformance to a single way of being. Water is still water whether it's a liquid, solid or vapor. Since the universe itself has undergone a variety of phase transitions, we may soon realize that it is also alive [Kaku 2005].

Like all healthy living systems, flourishing organizations know how to stay vibrant. They develop skills at:

- ✓ being deeply embedded in their context; thriving on interdependence; existing in relationship.
- ✓ embracing the truth of a situation; remaining open to all factors in the environment, not in denial of any; being neither naïve nor afraid of what they must face; coping with reality.
- ✓ rolling with changes in the external context; being immanently adaptable to the conditions they confront; experiencing quantum shifts, time and again.
- ✓ providing love, affection, belonging and support to their members; encouraging connectivity as a core strength.
- ✓ recognizing and trusting in inherent order and self organization; holding a few simple principles and being open to contingency and spontaneity; developing through learning and experimentation.
- ✓ renewing continuously in a way that maintains integrity; not having to manifest any one form or stance to retain integrity.

Living systems that possess these qualities survive and thrive. They are complex and extremely hard to extinguish. Those that don't, are short-lived and extinguish.

> Companies die because their managers focus on the economic activity of producting goods and services, and they forget that their organization's true nature is that of a community of humans.
>
> *Arie de Geus, 1997*

Introduction

💡 Approaches to Change

Unlike other species, our thoughts and actions are not limited to instinctual behavior. We can visualize ourselves in the future. In fact, we are continually considering alternative futures – whether we are planning dinner, choosing a career or selecting our company's next CEO. Part of living is to conceptualize and reinvent ourselves, our organizations and our societies based on the circumstances we expect to confront and the objectives we want to achieve. Our organizations morph from one incarnation to another in response to our anticipations.

Individually and collectively, we seem to have four basic approaches to change, each of which has its own logic:

- **Fix it:** *Alleviate the points of pain; solve the problem; return to the good old days; do more of the same ["It always worked before!"]; try harder, work longer; do more with less, cut the fat; reorganize; wait it out; stick with the tried and true*

  We study reports of past results to gather clues to the future. We are comfortable and safe inside our familiar world where we feel we can control what's happening. Anything that would change that world can be deeply disturbing. If something goes wrong, we seek someone to blame. Problems that arise need to be fixed. We may be living inside a bubble, but it's amazing how little anything beyond that bubble is relevant to us.

- **Let it collapse:** *Crisis gets people's attention; if it doesn't kill us, it will make us stronger, healthier, more resilient and better able to endure*

  In the face of serious problems, many believe we have to wait for a crisis to occur before we can take meaningful action. Indeed, crisis can be a wake-up call and lead to real change. Yet, we may not be in touch with the full implications of the incredible hardships the statement "It takes a crisis to get things moving" casually accepts. In crisis, we act quickly, out of fear, with no time to consider long-term implications. It is the most perilous and difficult time to change. We are in panic mode. It is a matter of survival!

- **Try something new:** *The world is changing; take a new approach; think outside the box; experiment; see what sticks*

  When we recognize that the changes occurring in our personal, organizational and natural environments are real, durable and have substantial implications, we see new opportunities as well as new threats. We let go of what was familiar and take steps to adapt to the changes with new approaches: new technologies, leadership styles, organizational structures and modes of work. Each step may be incremental, but the steps can aggregate to create profound change.

- **Leap into a new reality:** *Look at the world holistically; step into the prevailing energy; find latent forces ready to break through; direct momentum toward preferred outcomes*

  This path is familiar to startup organizations. Many outstandingly successful pioneers have created a new reality, even though the majority who took the chance failed to catch the next wave. For mature organizations, leaping into a new reality requires high levels of concentration and commitment; it is the most challenging, and potentially the most rewarding, of all four approaches to change.

Each of these orientations toward change is legitimate, grounded in rationality and historical validity. Organizational destinies aren't fated. We use foresight to explore plausible future

Introduction

conditions – both the good and the bad. We see what it will take to fix a situation so that we might continue as we are. We try new approaches and explore new realities to discover both expected and unintended outcomes. We are sometimes at our best when the chips are down. We can also drive change by acting in anticipation of a crisis while we have time to think clearly, consider options, test results and take sustainable action. We don't have to experience personal, societal or organizational crises *if* we can anticipate them and take effective, timely action.

When to Change

Good times can lull us into a sense of false security. When our organization is growing nicely, all we see is blue sky ahead; the wind is at our back! We fix any problems that arise and keep on going. We've found the key to success.

As an organization matures, limits to its growth will inevitably appear. If the demand for its goods and services grows faster than its ability to keep pace, the constraint is likely to be resource shortages. When this happens, things begin to look a bit different, somewhat less rosy.

This is exactly the right time to prepare for the next growth curve!

...before we find ourselves peering into the chasm of decline and failure, acting in panic and desperation.

Organizations that have the best chance to survive and thrive under fast-moving conditions are those that know how to evolve and adapt to change without losing their identity.
These organizations stand for something that endures. They evolve in pursuit of a compelling purpose, whose precise definition may organically transform over time. Such organizations have a magnetic quality, attracting and nurturing the talented people they need to adapt and grow, people who attract others like themselves. We envision a world in which these life-sustaining organizations flourish.

Introduction

Having run a company thorough a major transition, it's a lot easier to change when you can than when you have to. The cost is less. You have more time. I am a little worried that by the time we wake up to the crisis we will be in the abyss.

Paul Otellini,
CEO Intel, 2005 - present

Life-Sustaining Organizations

Life-sustaining organizations survive and thrive by continuously evolving in ways that affirm life. They put an explicit priority on the emotional and psychological needs of their workforce. They make a positive contribution to the world. They do no harm to the natural environment and, whenever possible, ameliorate the damage that may already exist.

Life-sustaining organizations have five special characteristics:
- ✓ Creative People
- ✓ Whole Systems Thinking
- ✓ Design Integrity
- ✓ Elegant Solutions
- ✓ Results Orientation

There is no inevitability to becoming life-sustaining. It takes determination, discipline and commitment to achieve lasting systemic change. This Guide lays out the steps that an organization can use to understand its external environment to consciously prepare for the future. This approach will support the life-sustaining qualities of any organization.

Creative People

> We are now at a point where we must educate our children in what no one knew yesterday, and prepare our schools for what no one knows yet.
>
> *Margaret Mead*
> *1901-1978*

No matter what the nature of the organization, its size or the scope of its activities, all work systems depend on talented, competent individuals. The faster the pace of change, the more dependent organizations are on securing the men and women with the skills they need to thrive under dynamically evolving conditions. Attracting and retaining this workforce requires a rethinking of traditional work environments, processes and tools.

The quality of the work experience can be a make-or-break factor for organizations in heated competition for key talent. A high-quality work experience allows people:
- ▸ to understand how their work fits into the purpose of the organization.
- ▸ to engage with others in a socially cohesive and productive manner.
- ▸ to achieve the satisfaction of a job well done because they have control and responsibility for assignments from start to finish.

Introduction

Organizations need to question their tacit assumptions about hiring, supporting, managing and retaining people to be sure that they remain employee-focused. Discovering and holding an employee perspective is critical to attracting and retaining fundamentally important talent.

Population growth is expected to slow over the coming decade, and consequently, labor force growth will also slow...The slower growth in the labor force relative to the population growth is a reflection of the declining overall participation rate ...driven by the aging baby-boom cohort as they enter the 55-and-older age group with a significantly lower participation rate than the 25- to 54-year-old age group.

Immigration plays a significant role in the dynamics of population and labor force growth. The Census Bureau estimate of population change between April 2000 and July 2006...43 percent was supplied by net international migration. Asians and Hispanics have been the fastest growing components of the labor force since 1986, and this is expected to continue into the future.

Bureau of Labor Statistics, 2006

If competition for creative talent increases – as numerous signs indicate it will – organizations will need the skills of a diverse range of individuals who are likely to be widely distributed geographically. In Europe, Russia, the U.S., and Japan, the number of talented people leaving the workforce greatly exceeds the number entering it. In spite of the current high unemployment rate, we expect to see a shortage of skilled workers in the not too distant future.

Life-sustaining organizations attract, house and support the creative talent they need to grow and flourish in the complex global context. Realizing that they are deeply dependent on their workforce, these organizations are inventive in identifying and recruiting the talent they need. They tap nontraditional sources in new ways and recognize talent wherever it exists.

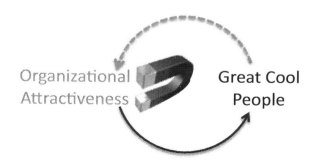

Great, cool people attract great, cool people. While "cool" isn't a precise term, we are using it loosely to signify Richard Florida's [2002] "creative class" or Daniel Pink's [2005] citizens of the "conceptual age." These are people who stand out in their fields. They are the ones others want to hang out with. Their styles define the look and feel of the organization. They are the center of "the scene." They are "what's happening." They may or may not be seen as "cool" outside their organization or industry, but they are definitely the buzz inside of it.

A life-sustaining organization becomes a magnetic attractor of dynamic, focused, "cool" talent. Strong talent in an organization is a powerful magnetic attractor. The more there is, the more there will be – being cool is a self-reinforcing loop.

Introduction

☐ Whole Systems Thinking

Comprehending a system is similar to the way we experience a digital photograph; although it is comprised of pixels, we see it as an information-rich image. Systems thinking is the art of being able to see apparently distinct events as interconnected and interacting. By viewing systems and learning how they function, we see the "big picture." By stepping back to view events as part of a whole, we can better understand why things happen as they do.

When events repeat over time, they form patterns. These patterns are driven by deep, underlying structures that shape the behavior of the system and have a profound impact on the behavior of the people within the system.

Life-sustaining organizations see systems at multiple levels:
* The personal
* The institutional
* The environmental

We need to understand the deep structures that drive life-affirming behavior at all three levels, because each depends on the other.

— The Whole Person

Life-sustaining work engages imagination, creativity and deeply held values. Most of us leave important aspects of ourselves behind when we go to work. Work seldom provides a place for all of our skills, interests, talents and enthusiasms; they are not welcome. For many, when we go to work, we deaden our real selves. We become covert: we display only those aspects of the self that we perceive to be welcomed and wanted. The rest has to wait for personal time. Here are a few snapshots:

- Ricardo worked for 20 years in a multinational corporation. He is now dedicating his efforts to founding a nonprofit organization. The idea of working for a large company again holds absolutely no appeal for him.

- Eric is a highly intelligent, entrepreneurial graduate student who cares deeply about people and the environment. From what he has observed of life in complex organizations, he will probably never join one.

- Janice, a Ph.D. in physics, loves her work, where she is creating something "the world has never seen before" while passionately managing a small group of professionals. She is also the mother of small children. Her energy, enthusiasm and competence seem boundless. However, because her values aren't represented by the organization's mission, she is actively seeking to make a career move to an organization whose principles and objectives are more aligned with her own.

Introduction

> - Piet, a dedicated young professional involved in improving housing in his country, shared an axiom by Eric Gill that guides his life: *"The artist is not a different kind of person, but every person is a different kind of artist."*

Work is an integral part of life: it is critical to our sense of accomplishment and well-being. It occupies a large portion of our waking [and dreaming!] lives. We lose some of our vitality and integrity if our work is out of alignment with our values and aspirations.

As individuals, we need to express our uniqueness. Each of us strives to think and act authentically, pursue our own destiny and differentiate ourselves from others. At the same time, we need a sense of belonging and connection; we want to feel recognized, accepted and appreciated. We need to be part of a supportive community. We need to have our independence and to be team players, simultaneously. The artist within each of us seeks to compose a life in harmony, beauty and balance.

We have been studying festivals to understand why people, for several months every year, demonstrate a significantly higher level of passionate engagement in the preparation for their festivals than they offer in their workplaces. The members of festival groups are transformed through their passionate engagement in a collaborative, creative design process and the ritual of becoming a member of a group, where their identity and success are closely linked to the identity and success of the group.

Michael Diggiss and Roosevelt Finlayson
The Caribbean Centre for Quality and People

A life-sustaining organization seeks to satisfy these disparate needs by working with the people of the organization to create a shared vision. The organization welcomes the whole person into the workplace and helps each individual reach his or her full potential. This support enables workers to elaborate their individuality within the context of the organization's identity and purpose. A life-sustaining organization evolves in ways that align with the aspirations of its employees and, by so doing, sustains its own growth and economic vitality.

Many women are leaving high-paying positions to start their own companies [Lesonsky 2010]. They are doing so to create a high-quality work experience that has eluded them. They are willing to work long hours for low pay to be whole. If organizations want these sorts of self-starters, they are going to have to work harder to get and keep them!

At the Forum on Workplace Flexibility 2010, President Obama said, "What is good for your workers is good for the bottom line."

The Colours Junkanoo Group, lead by Christian Justilien, performing at the 2nd International Dialogue on Festival in the Workplace (FITW) in Nassau, The Bahamas. See www. bahamasentertainers.com

Introduction

— The Whole Organization

Life-sustaining organizations rely on people to care both about their own work *and* the organization as a system. People work best when they have pride in the organization's mission, goals and strategies. Each subsystem of the organization is a fractal of the whole, embodying all the elements of the whole. Every person in the organization needs to grasp the fact that they are in this together, that a problem in one area affects them all. They need to understand how their individual and collective efforts are tied to something larger. Thinking together for the common good of the organization and having the big picture in mind helps people act in ways that align the organization around its common bonds and shared aspirations. Systems thinker Barry Oshry uses the word "partnership" to describe organizations in this sort of holistic state, where every element of the organizational system manifests the whole appropriately and fully in the context of local conditions. Each discrete part functions harmoniously as part of the whole organism. When operating in partnership, the people and the organizational functions are committed to their mutual success and to the success of the organization as a totality. Real partnership is an emotional state that sustains the life of each person within the system and that of the organization as a whole [Oshry 1999, Sales 2006].

In the world of living things, every system can be more real or less real, more true to itself or less true to itself. It cannot become more true to itself by copying any externally imposed criterion of what it ought to be. An organization is whole according to how free it is of inner contradictions. This subtle and complex freedom from inner contradiction is just the very quality which makes things live.

Christopher Alexander, 2002

— The Whole Planet

Because the life-sustaining organization is aware of the larger context within which it exists, it is infused with a deep feeling for its effects on its external environment, locally and globally. It is a cell of a living organism called Earth. Living systems constantly renew themselves in ways that maintain the integrity of their structures while acknowledging that they are deeply embedded in and shaped by all aspects of their environmental context, e.g., historical, economic, political, emotional, spatial and technological [Jantsch 1980]. A life-sustaining organization is very much part of something, yet it is fully alive as a vibrant, inimitable self. It is as natural for this kind of enterprise to function in harmony with its environment as it is for a bee to pollinate flowers as it gathers nectar to make honey.

Today's challenges are so great — and the dangers of the misuse of technology are so global, entailing a potential catastrophe for all humankind — that I feel we need a moral compass...a holistic and integrated [recognition of] the fundamentally interconnected nature of all living beings and their environment...We need to relate to the challenges we face as a single human family...Some might object that this is unrealistic. But what other option do we have?

The Dalai Lama

Introduction

We will use the term "environment" at three levels: the work environment of the individual, the business environment of the organization and the natural environment of the Earth. To help distinguish among these, we will adopt the following nomenclature:

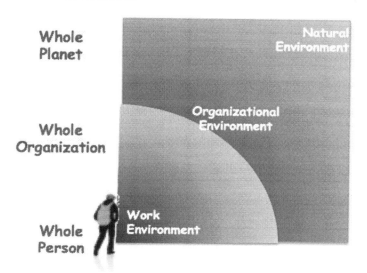

Natural environment
The global environment, including the social, political, and economic forces that affect nature.

Organizational environment
An organization's external environment that directly impacts its functioning – its customers, markets, competitors, suppliers, partners, etc. The organization's context.

Work environment
An individual's experience at work, including virtual and physical space, tools, culture, atmosphere, expectations, etc. We also refer to the work environment as the workplace.

☐ **Design Integrity**
People are receptive to and inspired by the power of design, and nowhere is this more true than in their response to their physical context at work. Work environments are intended to enable people to do their jobs, and yet office design has become largely soul destroying.

> The work environments that organizations have been providing are increasingly unsuited to emerging patterns of work and are inhibiting workers from performing to their full potential.
>
> *Audrey Schriefer, 2005*

The repetitive, modular design of many workplaces caters to space and cost savings more than to workers' needs and the most effective way to get the job done. We try to personalize workspaces to our own preferences, but the results are insufficient. Any connection to personal life, family, aesthetic tastes or community has been sacrificed and depersonalized. Perhaps this is one reason why so many people jump at the opportunity to work anywhere other than the corporate office.

There is no universal best workplace design. The workplace must embrace the uniqueness of the organization as well as its individual employees. Work environments designed to truly facilitate work allow more of the whole person to emerge.

The work environment includes both tangible and intangible aspects of the workplace – the look and feel of the work experience. In life-sustaining work environments:

Introduction

- The *look* of the physical environment supports the comfort and safety of individuals and simultaneously feeds their sense of curiosity, beauty and creative expression. Its characteristics consistently and authentically express the values and uniqueness of the organization – where it came from, what it stands for and where it is going. It provides a positive sense of place that allows the imagination to soar.
- The *feel* of the work environment includes:
  - A sense of community that energizes and engages
  - The functionality of the furnishings, equipment and technological tools
  - Policies that facilitate, rather than impede, productivity
  - Leadership and peer attitudes of respect and trust
  - The culture and atmosphere

When the elements of the work environment truly support work and the worker, employees are fully engaged and social networks are alive and well. The strength of a life-sustaining environment comes from the mutually enhancing interplay of...
- the people of the organization
- the places they work
- the processes they use to get things done.

Forward-thinking organizations of all sizes and across all industries recognize that elegantly designed and innovative workplaces can enhance employee and business performance, resulting in long-term cost savings and improved organizational results. These companies have tailored their workplaces to meet their particular needs and those of their employees. They use their space and technology investments to enable, rather than inhibit, progress toward their objectives [Schriefer 2005].

Telework is another powerful emerging force that allows the people and organizations to craft new patterns of interconnectivity between work, home and community. The multifaceted significance of remote work is driving many innovations in the shape of the work environment. Life-sustaining organizations embrace both physical and virtual presence in various and changing combinations to support their workforce in being fulfilled and productive people.

☐ Elegant Solutions

A life-sustaining organization functions in ways that allows it to enduringly fulfill its highest purpose. It does this simply and consistently, allowing a great deal of autonomy to its employees. As an example of how this has been accomplished in an urban context, consider Georges-Eugène Haussmann's legacy following his renovation [some

Introduction

would say creative destruction] of Paris in the 1860s. He imposed three rules on the city that endure to this day:
- A uniform building height
- A restricted palette of materials
- The requirement to wash building facades every 10 years

From these few rules to guide actions, a city of great beauty, infinite variety and ordered complexity emerged and continues to evolve. The intricacy of the patterns and design details is infinitely richer than any one designer could have produced. By limiting the options, a consistent and pleasing palette of solutions blossomed.

The idea of creating complexity from simplicity and a high level of autonomy is demonstrated mathematically in "chaos games" [Barnsley 1993]. The result is a fractal image like the Sierpinski triangle...

The rules of this game are:
1. Inside a large triangle, place a smaller one with a base parallel to the horizontal axis.
2. Make a copy of the inner triangle, shrink it to half the height and width, make two additional copies and position them so that each touches the sides of the larger triangles at a corner.
3. Repeat step 2 with each triangle.

An "emergent property" of ever greater complexity and intricacy manifests from the repeated iterations of simple rules.

Work environments that learn from and mimic natural phenomena put in place a few principles that resolve complex problems with astounding variety and overall cohesion. These variations on a theme are intricate and extremely subtle.

Life-sustaining organizations integrate a number of systems, including workflow processes, workplace design, productivity and communication tools, organizational policies, structure, norms, etc. What we might call "integrative leaders" understand how the many components of an organizational system are intended to support one another in a cohesive whole. By grasping the relationships among these

Introduction

elements, leaders find that they don't need to focus on each factor in isolation. Processes of adaptation continuously go on.

In *Whispering Pond*, Ervin László [1996] holds that our every thought, feeling and relationship resonates widely, instantly and eternally. All contribute to a subtle universal field that is the infrastructure of psychic energy for matter, life and mind. Solutions are shaped by the person, the group and the function creating and applying them. As we deepen our understanding of the forces that drive a system, we can simplify our interventions with the assurance they will have the intended effect. Elegant solutions emerge. In an organization that has creative people operating with a whole systems perspective, leaders can apply simple rules that allow employees a great deal of situational independence. Thus, elegant solutions point individuals and groups toward interdependent, but autonomous, action.

The patterns which control a portion of the world are themselves fairly simple. But when they interact, they create slightly different overall configurations at every place. This is the character of nature.

Christopher Alexander, 1979

☐ Results Orientation

As organisms, life-sustaining organizations compete for needed resources – to survive and, what is more important, to thrive. The fact that their people are attuned to changes in their internal and external environments gives them an advantage. They are infused with an attitude that anything is possible. They are curious and relish innovation, balancing passion with responsibility. They are risk takers, not afraid of making mistakes. Learning enables organizations to produce results. Hiding from mistakes is the opposite of learning.

Life-sustaining organizations are not judgmental. Because they are attuned to what's happening, they can take advantage of opportunities and respond to threats much faster and more effectively than organizations that are blindsided by unfolding events. By achieving sometimes astonishing results, they shape the context they are in and in turn are shaped by it. For example, Google's confidence in its ability to influence its organizational environment provides a solid basis for its willingness to challenge the dictates of China's totalitarian government. Other powerful companies seem unwilling to take the same risk. Google's encounter with China has heightened its attractiveness to a group of extraordinarily talented people [Jacobs 2010]. Life-sustaining organizations get results and raise the bar for everyone else.

These forward-thinking organizations attract a cadre of creative people, encourage them to practice whole systems thinking and design integrity in all aspects of organizational life, search for elegant solutions and generate results, yielding a life-sustaining organization. Once an organization has become life-sustaining, it is in a much better position to draw and

Introduction

hold the creative talent it needs. Its core elements are regenerative and stable, like a fractal or a beautiful flower.

- **Your Experience**
  Have you been part of an organization that embraced some or all of these characteristics?
  → If so, visualize the place, the people, the activities, the impact of the organization on matters that you really cared about.
  → If not, imagine how it might have been.

> **Personal Encounters with Life-Sustaining Organizations**
> - ✓ Which members of the organization deeply affected your life? What made them so special?
> - ✓ Who was valued? Who was ignored, demoted, let go? Why?
> - ✓ Were there any features of the physical workplace that held particular importance to you? What were they and why were they meaningful?
> - ✓ Was the organization on the cutting edge in some way – technology, work processes, decision-making policies, social networking strategies, etc.?
> - ✓ Did this organization create some expectations or standards that you've carried into the rest of your working life?
> - ✓ Was there anything special about the way that you were trusted or the degree that you trusted others in this situation?
> - ✓ If you and others were particularly committed and engaged, what was it about the organization or the work that allowed that passion to flourish?
> - ✓ Was there something that you got to do in this organization that you never got to do anywhere else?

Jot down a few notes. You may want to reflect back on your personal experience as you continue through this Design Guide.

Growing a Life-Sustaining Organization

An organization exists within a larger context. As a living system, it responds to external dynamics in order to flourish in its complex environment. Organizations embodying a "fix it" or "let if collapse" approach to change may miss the optimal entry point into a change process [i.e., when things are still going well but you can see problems over the horizon]. An openness to what is happening in the organizational environment is fundamental to being a living organization. It provides immunity to complacency and negativism. Becoming consciously aware of itself as a living system is the best way for an organization to become adept at making necessary changes.

If an understanding of its external environment is so important, how does an organization discover what really matters to its future amid the enormity of contradictory events occurring continuously – both promising and discouraging? How can an organization hold a shared vision that each person in the organization truly believes in and actively implements?

Introduction

In this Guide, we offer two pathways that reinforce one another – the Structural Dynamics approach to strategy development and implementation and Anticipatory Leadership to create bench strength at every level of the organization.

- **Structural Dynamics**

  This Guide describes a rigorous process to generate insights about what is going on in an organization's external environment, why it is happening and what it means to the organization at an operational level. Simultaneously, the process builds internal capacity to face the unexpected and make the dramatic changes the analysis frequently demands. The process in itself contributes to the learning of the organization that undertakes it and develops leadership capacity throughout.

  We use the word *structure* to signify a stable framework, like the human skeleton, which both shapes and supports the functions and activities of the system. The word *dynamic* indicates that the state of the system varies widely – from equilibrium to metamorphosis, and everything in between.

  This Guide steps through the components of Structural Dynamics. We start by:

  ▸ **Convening** a team of strategic thinkers from all parts of the organization

  The learning process includes these phases:

  ▸ **Exploring** facts – the system of forces interacting to shape the future

  ▸ **Discovering** options – leveraging the facts for strategic advantage

  ▸ **Embodying** actions – embracing the learning throughout the organization

  ▸ **Sustaining** results – moving forward with confidence while maintaining and transitioning current operations.

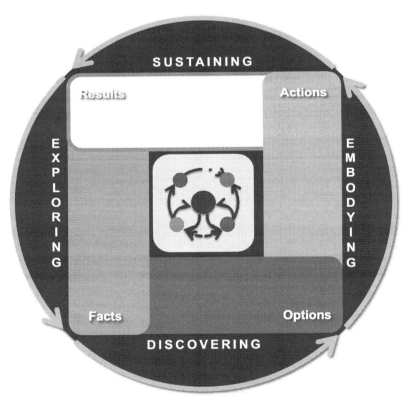

Structural Dynamics can be used for any complex strategic issue that confronts an organization. Using this approach develops Anticipatory Leaders and Learning Organizations. In this Guide, we are describing an application of Structural Dynamics to the strategic issue of developing life-sustaining organizations.

Introduction

- **Anticipatory Leadership**
  Participants in a Structural Dynamics process engage in a team learning experience while developing a shared vision for their organization. They also hone essential leadership skills required in our increasingly complex, fast-paced world. These "Anticipatory Leadership" skills become second nature as participants practice them with one another and model them in their role as communicators throughout the organization. Anticipatory leaders are:

  Integrators who bring people together for strategic dialogue and action,

  Futurists who encourage a broad exploration of the forces shaping and reshaping the context of the organization,

  Strategists who orchestrate creativity in the development of strategic options and their implentation.

  You will learn about the role of the Anticipatory Leader in each phase of the Structural Dynamics process. The conceptual thread of Anticipatory Leadership is developed throughout this Guide, along with the specific concepts and activities that characterize each phase of the work and the process through which all members of the organization, from senior executives through front-line employees, are engaged. Not only do they practice these skills in the process, they concurrently apply them in their interactions with others throughout the organization. Thus, Structural Dynamics is a process that develops a broad base of Anticipatory Leaders.

  Anticipatory leaders, using Structural Dynamics, facilitate organizational learning. Structural Dynamics ingrains habits of investigation, reflection, discussion and experimentation into all aspects of organizational activity and sensitizes the organization to the impact of external forces. This greater situational awareness enhances the organization's timely, effective responsiveness.

How to Use This Design Guide

This Guide is intended for everyone who wants their actions to shape their organization and its future. These include senior executives, line managers, those with specific creative skills [e.g., scientists, artists, designers and craftspeople], technologists, specialists, strategic thinkers and network leaders. Organizational directors, officers and their advisors will find this book to be of value in their roles. Academics and students of organizational science will find value in this approach as well.

This Guide shows how to use the Structural Dynamics approach and Anticipatory Leadership skills to create life-sustaining organizations. The Guide has two parts: you will learn about each phase in Part 1; Part 2 illustrates their application.

- **Part 1: Becoming Life-Sustaining**
  Each chapter in Part 1 starts with a description of the Anticipatory Leadership skills that are most needed at each phase of the process. The chapter goes on to describe the theory and application of Structural Dynamics step by step.

Introduction

- ▸ ***Convening*** brings together a group of diverse, influential people to take on the challenge of redesigning the organization to be more life-sustaining. To convene this strategic team, leaders practice their skills as integrators.
- ▸ ***Exploring*** looks at the organization's external environment to gain an understanding of the forces in play and to model them in a way that enables visualization and, thus, insight. Exploring facts relevant to an emerging future, leaders develop their skills as futurists.
- ▸ ***Discovering*** creates scenarios of future possibilities in which the team can develop and stress-test strategies. The most robust strategies, those that work in a range of future conditions, emerge from the process. To discover their most effective options, leaders exercise their skills as strategists.
- ▸ ***Embodying*** uses a living systems approach to embed strategy into operations throughout the organization. To embody the strategic options, leaders again rely on their integrative skills.
- ▸ ***Sustaining*** nourishes the completed cycle of action to strengthen the organization's capacity as a learning system. This phase calls upon all the skills of Anticipatory Leaders.

- **Part 2: A Case Study**
  The composite case depicts the application of Structural Dynamics and Anticipatory Leadership in some detail. Each organization is unique; the process shapes and molds itself to fit particular situations.

This Design Guide can be read in any of the following ways:
1. Overview: To get the essence of this guide, our core ideas and our values, read the Preface, the Introduction and Final Word.
2. Cover to cover: Read the book from start to finish to get a complete picture of what makes an organization life-sustaining, why your organization could benefit from becoming life-sustaining, and how to turn on the life-sustaining power of the organization you care about.
3. With a focus on Anticipatory Leadership: Use the sections at the beginning of each chapter in Part 1 to enhance your skills as an Anticipatory Leader.
4. With a focus on the Life-Sustaining Organization: Start with the case study in Part 2 to learn how an organization becomes more life-sustaining and what results it can achieve.
5. With a focus on the Structural Dynamics strategic leadership process: Every step is richly described in Part 1 and illustrated in the case in Part 2.

Enjoy!

If you have questions or comments, contact us at info@artofthethefuture.com.

Part 1: Becoming Life-Sustaining

Convening The Team

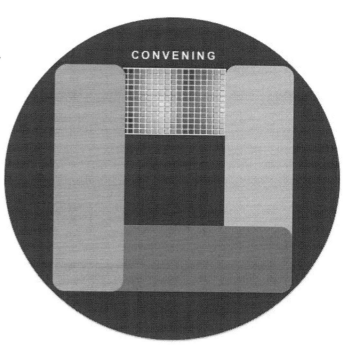

- **Role of Leaders: Integrator**
  - Integrating Creative Talent
  - Engaging in Whole Systems Thinking
  - Expressing Design Integrity
  - Seeking Elegant Solutions
  - Orienting Toward Results
  - Your Experience
- **Preparing for the Life-Sustaining Initiative**
  - Choosing a Facilitator
  - Assembling a Powerful Team
  - Documenting and Communicating – Not Representing
  - Modeling a Life-Sustaining Work Environment
- **Convening**
  - Articulating the Decision Issue
  - Selecting a Time Horizon
  - Establishing Criteria for the Playbook
  - Making Agreements and Setting Expectations

Convening

◉ Role of Leaders: Integrator

As the twenty-first century gains traction, the future beckons as a multifaceted field of potential energy. In a frictionless environment, the future could go in any direction. Of course, we do not exist in a frictionless environment; many forces are in play shaping the future. Most people have ideas – some very strongly held – about the shape of the future. Individual perspectives rarely grasp all the possible variations of future possibility. The deluge of information bombarding us from an ever-expanding set of disciplines and subspecialties is beyond the ability of any individual to comprehend. We tend to focus either on a specific subset of forces – or on the strength of inertia keeping things in place. When dominant voices are given too much credence, other people with equal or greater insight may be drowned out and go unnoticed. But all members of an organization have views on what is enhancing or limiting its life-sustaining qualities. They need to be encouraged to share what they know and what they sense about the forces shaping the organization's destiny.

Leaders manifest a strength of character and presence that commands the attention of others. When they realize that their organization is a living system and commit to helping that system become life-sustaining, something significant occurs. Such leaders learn to use their positions to create cohesive teams and access the full potential of each individual in the organization. They engage others in passionately and persistently supporting the purpose and strategic direction of the firm. They are able to do so because the workforce cares deeply about the emergent life-sustaining organization.

Leadership exists everywhere within the organizational structure. Not all leaders are in formal positions of authority. Even in "flattened" organizations, formal and informal leadership endures. In this Guide, we refer to three types of leaders. Organizational health and success depends on all three types acting in concert ...

> Leadership is too important to leave to the CEO.
>
> *Peter Senge, 2009*

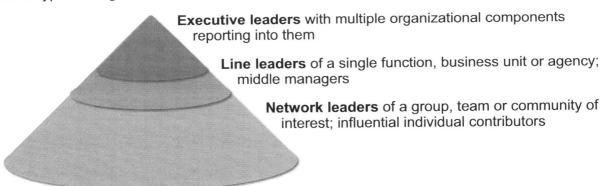

Executive leaders with multiple organizational components reporting into them

Line leaders of a single function, business unit or agency; middle managers

Network leaders of a group, team or community of interest; influential individual contributors

- **Integrating Creative Talent**
  Whatever their organizational position, integrative leaders possess an ability to bring people together across organizational boundaries. In doing so, they overcome the forces of inertia that limit organizational achievement when people work solely within their "stovepipes" or "silos." The perspectives of various functional groups differ from one another and sometimes clash [e.g., marketing frequently cannot understand engineering and visa versa]. Integrative leaders promote dialogue and cross-functional initiatives that bring people into discourse and mutually supportive action. They encourage collaboration and knowledge sharing to discover new possibilities. They thrive when individuals and the organization perform at their highest level. By encouraging people to engage in conversation and work together, integrative leaders bring the whole living system into alignment.

  Integrative leaders engage people's hearts and minds. They recognize talent in people of all ages, genders, ethnicities, religions and nationalities. Through their listening skills and empathy, integrators nourish people. In considering the thoughts and feelings of others along with their own, integrative leaders display genuine respect. Simultaneously, they influence others to do the same. These behaviors enhance knowledge sharing and organizational effectiveness. Once an organization has established a reputation as a great place to work, people flock to it. As a result, a life-sustaining organization is made up of great, cool people with tremendous commitment to the organization's mission and culture.

- **Engaging In Whole Systems Thinking**
  Every function and operation within a life-sustaining organization shares ownership in the future of the whole system. It is impossible to have a completely healthy organism if each organ is not supporting all the others. However, there are many forces at work in complex systems that make it difficult to achieve this alignment of purpose. Unless consciously addressed, the constituent parts of an organization will naturally seek to optimize their own results often to the detriment of the performance of the whole. Here's how fractionalized thinking works at the line leadership level:[1]

  1. Organizational growth in size and complexity requires specialization.
     - In startup organizations, everyone does everything.
     - As an organization becomes larger and more complex, specialization is required.
     - The larger the organization grows, the more specialization it needs.

---

[1] This section draws on the work of Barry Oshry.

Convening

2. With greater specialization and complexity, more demands are placed on line leaders within each specialization.
   - Their attention is pulled between the demands of executive leaders and the needs of their staffs.
   - Line leaders have little time to interact with their peers, sharing knowledge and addressing joint issues.
   - Line leaders' focus is narrowed to their own "report card;" they each see their part and neglect the needs of the whole.

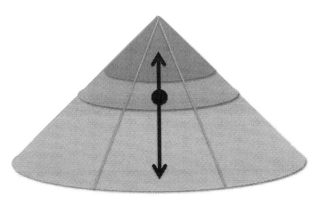

3. This separation creates emotional and intellectual distance between parts of the organization.
   - Each organizational component, with the best of intentions, seeks to optimize its own potential.
   - The sense of what is best for the living system as a whole is degraded.
4. Without strong ties to one another, the line leaders develop a sense of alienation from their peers.
   - Line managers start feuding and blaming.
   - Systemic issues are addressed as "personality conflicts."
   - The system become disconnected and fragmented.
   - Organizational results suffer.

Increasing complexity, specialization and narrowing of focus also cause leaders in these fragmented systems to think only operationally; long-term, strategic thinking is put "on hold" for another time that never seems to arrive. We all apply long-range thinking to many aspects of our personal lives outside of work – choosing careers, homes and partners; planning families and lifestyles; saving for education and retirement; and anticipating the legacy we'll pass on to future generations. However, working life frequently encourages narrow, short-term thinking, and, therefore, our natural strategic thinking abilities often atrophy [Argyris 1957].

Executive leaders with their own specialized responsibilities often remain unaware of the forces that isolate line leaders and don't see the need to drive integrative processes. Over time, this separation creates "stovepipes" that become permanent features of organizational culture. Their many debilitating effects include internal competition, interpersonal conflicts and mediocre performance of the entity as a totality. There is neither joy nor sustainable achievement in this picture!

Convening

Integrative leaders see the organization as a whole rather than concentrating exclusively on their own domain. They build partnerships to aid the organic evolution of the whole system rather than disconnected, impotent fragments.

To counter the systemic forces that cause line leaders to disperse, they must keep the channels of communication open and encourage respect and cooperation between organizational components. Executive leaders who are adept integrators support multiple means of communicating across organizational boundaries. They acknowledge and reinforce behaviors that benefit the whole as they happen. Such interactions allow line leaders to:

- ✓ Develop an awareness of internal organizational problems and opportunities.
- ✓ Contribute their perspective and expertise to the whole.
- ✓ Participate wholeheartedly in the implementation of organization-wide initiatives.

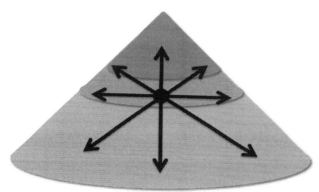

These interactions maximize the benefits that come from specialization by creating a culture of information sharing, joint problem solving and collaboration among line leaders and their staffs. Once they have the opportunity to work collaboratively, they will discover that they can be helpful to each other in many ways. Actions that unify people at all levels enhance the vitality of life-sustaining organizations.

- **Expressing Design Integrity**

    The look and feel of a life-sustaining organization is aesthetically cohesive. Nowhere is the presence of an integral design sensibility more important than in the work environment. A life-sustaining work environment supports, rather than inhibits, people striving to perform to their full potential. When people feel good about where they work, what they do, and the interactions they experience with peers, leaders, staff and customers, the payoff to the organization is enormous. The character of the physical space and the quality of the equipment and technology they use completes the sense of wholeness, cohesion and satisfaction. Consciously cultivating such environments requires integrative leadership.

    The aesthetics, culture and comfort of the work environment and the facilitation of work affect everyone in the organization. Responsibility for shaping it involves many internal functions including real estate, facilities, human resources, information technology, business development, strategic planning, organizational development, finance and the user group that will be working in the environment that is being created or redesigned. Typically, people in each of these functional groups fill their roles as best they can, but

Convening

the work environment as a whole is not integrated. No one is approaching this crucial aspect of organizational well-being comprehensively because it does not fit neatly within a single domain. Concern for the design integrity of the work environment falls through the cracks. As a result, the workplace often becomes a hinderance to peak performance rather than the facilitator and the organization sacrifices tremendous potential.

An integrated work environment reinforces an organization's unique culture, purpose and strategic intent. A cohesive look and feel aligns the organization by unambiguously broadcasting a compelling message. Integrative leaders' attention to this often missed opportunity can yield life-sustaining results.

- **Seeking Elegant Solutions**
  In the search for simple, elegant solutions, the integrator allows insight to emerge. Rather than a cumbersome set of rules and guidelines to cover every conceivable condition, a few memorable ones capture the imagination. They engage the capability and the integrity of the workforce to apply them in the intended spirit. To encourage cross-functional dialogue from which elegant solutions emerge, integrators convene people involved in the issue. It may be more difficult to let an elegant solution emerge than to prescribe every step, but the rewards include more understanding and cooperation among the various parts of the organization and more engagement of every member of the organization in the organic application of the solution.

- **Orienting Toward Results**
  A cross-functional dialogue is a forum to grapple with complex issues and search for elegant solutions. Those involved develop personal ties and a respect that enables them to broach a range of sensitive and important topics. The search for elegant solutions in one area may well foster organizational growth in others. For example, a conversation about work norms touches on many aspects of organizational life. It will almost certainly lead to the discovery of connections between the norms and policies of the organization as a whole and the nature of the work experience. It will reveal interconnections between the work environment, the organizational environment and the natural environment. Just as a flower blossoms because that's what it does, a strategic dialogue that is well convened and nurtured bears results – which may be unexpectedly stunning.

- **Your Experience**
  To evaluate your skills as an integrative leader, take a few moments to consider the following. We don't propose this as an exhaustive assessment, but the skills we inventory below are those that we've noticed among integrators. This brief survey may highlight strengths already present in your way of thinking and leading.

Convening

> **Integrator Self-Assessment Tool**
>
> ☐ I reframe the thinking of others so that they become aware of the possibilities in a situation.
> ☐ I seek, value and take advantage of all the good ideas and positive, enthusiastic energy in a system.
> ☐ I am at ease coaching and being coached; I am comfortable seeking feedback from others and incorporating it into my behavior and attitudes.
> ☐ I show people where their points of agreement and commonality lie.
> ☐ I am skillful at facilitating the efforts of others so that they arrive at desired results and know that they have grown as a result of the process.
> ☐ I possess a positive vision.
> ☐ I bring people into alignment with my vision by demonstrating how it moves them and the organization ahead.
> ☐ I am accessible and highly responsive to other people.
> ☐ I value others' contributions, and find and openly admire qualities in others.
> ☐ I maintain awareness of the good intentions of others.
> ☐ I am open to learning from everyone, regardless of their organizational level, function, social position or degree of recognition.
> ☐ I take satisfaction in the successes of others.

You may want to elaborate your response by putting a numeric value in the box [e.g., "5" if you manifest this behavior or attitude consistently and successfully] and by identifying one to three detailed examples of your demonstration of the skill in question or of the difficulty you've had with it. If you have demonstrated strengths in some of these domains but less so in others, you may want to look for opportunities to practice particular competencies.

🌐 Preparing for the Life-Sustaining Initiative

The success of a life-sustaining initiative can be undermined from the start if organizations do not pay attention to some essential considerations.

1. The executive leadership of the organization must be solidly in support of the initiative for it to have a chance. If the idea is initiated by network or line leaders, they must be sure to gain executive support and commitment.

Convening

2. The leader of the initiative must be a strong leader and respected player who is able to maintain the necessary support at every step.
3. Convening the right people to participate in the initiative is an art. The team must represent a wide range of views and opinions present within the organization.
4. All participants in, leaders and supporters of the initiative must take an active role in communicating throughout the organization and receiving information from start to finish. Electronic forms of communication are essential for scope, coverage and consistency but should not be relied upon to replace face-to-face dialogue.

We will refer to several roles throughout this Guide.

Sponsoring Executive – A power holder in the formal organization who can supply the resources to fund the initiative and the intellectual and emotional commitment to wholeheartedly support it all the way through implementation. This is usually the most senior executive of the organization. If a component part of the organization is undertaking the initiative, it is the senior leader of that group.

Team Leader – A skillful and influential member of the organization who shapes the Team and directs the process [This may be the same person as the sponsoring executive]. The Team leader is also an active participant in the process, a member of the Team.

Facilitator – A person trained and experienced in the specific skills of facilitation. This person manages the process, keeping it on track toward completion while encouraging the active participation of all Team members. He or she monitors people's adherence to the norms and behaviors they have agreed to and maintains a psychologically safe space, i.e., one where they feel comfortable doing their best and most creative thinking.

Team Members – People who care deeply about the organization and its future. As a group, they embody a broad range of perspectives. They agree to wholeheartedly participate in the process for the benefit of the organization as a whole. The size of the group varies between 16 to 32. We've found 24 to be a good size to get diverse perspectives – even in sub-groups– while ensuring that everyone has the opportunity for sufficient "air time" – so that all perspectives are heard and considered.

Playbook Coordinator – An information technology specialist with well-developed communication, knowledge management and design skills. This person is selected by the Team leader and is dedicated to the initiative throughout its duration. The coordinator supports the Team in electronically communicating with all employees.

- **Choosing a Facilitator**
  Trying to play two roles on the team simultaneously – facilitator and participant – is usually impossible to do well [Sales and Shuman 2002]. Because it is important for every Team member's voice to be heard, the Team leader needs to participate. If the Team leader also facilitates the discussion, then he or she may inadvertently dominate the conversation. We strongly recommended that the organization select an experienced, trained facilitator to guide the process from the outset.

- **Assembling a Powerful Team**
  An organization that wants to thoroughly explore its future possibilities needs a Team of strategic thinkers who tend to look at the big picture, know what's going on in the organization's external environment and are able to discern facts from opinion. Growing

Convening

a life-sustaining organization means going beyond the obvious and considering possibilities that may be uncomfortable and unpleasant, even unthinkable.

Team members are drawn from various internal constituencies and can be supplemented by external players at the organization's discretion. Selecting the right Team makes the initiative representative of the full range of views existing in and around the organization. The objective is to engage the whole organizational system.

Although it is rarely practical or possible to have every member of a complex system physically "in the room," everyone's view can be heard. The Team leader, acting as an integrator, needs to think expansively about how to capture as many diverse views as possible. Holding this objective will lead to a Team that reflects the various functions, levels and interests within the organization.

The process thrives on a wide range of views based on facts knowledge and. A diversity of views is not the same thing as conflict. Team members will be both sharing their knowledge and learning from others. The balance of advocating for a stance and inquiring into the unknown or unconsidered is essential. It requires participants to respect the ideas and perspectives of their teammates. There are no right answers when learning from the future. Opposing points of view are validated and woven into distinct scenarios to explore a range of plausible future possibilities. A group composed of people with similar mindsets would find it difficult to consider a broad range of plausible futures and, thus, would weaken the analysis.

> Co-inquiry in the presence of...massive diversity almost always discloses views not quite like our own...we are often...surprised by the ending.
>
> *David Cooperrider, 2007*

The composition of the Team is a key element for a successful initiative. The goal is to get a strong mix of competence, sound knowledge, capacity and energy. Here are some criteria for assembling a Team:

- **Strategic Thinkers**
  People who think long term, beyond the boundaries of their own domain and even beyond the boundaries of the organization and industry. They absorb a wide range of information and consider the long-term implications.

- **Divergent Viewpoints**
  Widely differing views based on objective data spark the thought process and enliven the discussion. They ensure the consideration of a broad range of possibilities.

- **Active Listeners**
  The ability to hear and consider the views of others as well as advocate a position is a critically important characteristic.

- **Cross-Functional Players**
  The Team should be drawn from all parts of the organization.

- **Diversity**
  A mix of length of service, age, race, gender, ethnicity, physical ability, sexual orientation and education level helps to obtain a wide range of perspectives. However, quotas are strongly discouraged.

- **Multiple Levels**
  Leaders and independent contributors at all levels who are respected within the organization and able to make themselves heard. Be sure to include some front-line employees.

Convening

 Network Hubs
Influential men and women who are social networking communication hubs in the informal organizational structure, regardless of their formal positions.

The process of forming such a Team is a challenge. It is important to choose Team members in an inclusive way that goes beyond considering only the most obvious candidates. The candidates and their managers should clearly understand the amount of time Team members will be required to spend away from their regular jobs. The nominee for Team membership also must want to participate.

Here are three approaches for selecting members that we have seen succeed. A combination of these approaches or some other method entirely may be best suited to your organization.

a. Senior leaders of the organization meet to suggest candidates and sort through those named until a Team with the desired mix of characteristics emerges.

b. The Team leader solicits nominations for Team members from all parts of the organization. By consulting with colleagues, he or she seeks to learn more about the nominees and gradually composes a slate of candidates.

c. The Team leader and executive sponsor use their networks to identify candidates throughout the organization. They vet potential candidates with their managers and others in the organization who know their work, their character and their style.

Each organization needs to find an approach that works in its culture. The selection process itself may represent a shift in behavior toward becoming a life-sustaining organization. The process may not happen quickly or without a hitch; be forgiving and persistent.

The Team leader may want to interview candidates before inviting them to participate. Once a list of candidates for Team membership has been compiled, executive leaders communicate with the candidates' line managers, who in turn invite the candidates to participate. This invitational conversation should include a description of the selection process, the criteria that were applied [i.e., why this candidate was chosen] and the expectations of Team membership. Candidates have the right to accept or decline the nomination without any repercussions.

People from outside of the organization can also play a valuable role in the initiative. In most organizations, people rarely discuss strategic issues with outsiders. Yet it can be well worth the risk to get fresh perspectives on unquestioned assumptions, organizational myths and long-held beliefs. It is also a powerful way to find out what's going on beyond the organizational boundaries. Even if you think you already know, you may be surprised. Here are a few ideas for people the organization might invite to participate as Team members or as guest presenters early in the initiative to inspire expansive thinking:

 External Stakeholders
Organizations need to consider who has an interest in the organization's success. Key suppliers, for example, have reason to care and may have valuable information to bring to the process.

 Subject-Matter Experts
Experts such as futurists, technologists and workplace specialists have information [e.g., the results of relevant social science research], experience and competence

that could supercharge the strategic initiative. They can ensure that the Team is aware of recent or potential developments. Without a complete enough set of good data, the process could miss important possibilities. It is important to ensure that these experts present facts in any unbiased way with no hidden agenda.

Customers
People using the organization's goods and services [or those who should be] are great sources of insight into the forces affecting current and future decisions.

If external players are not included, the Team leader must be certain to obtain the knowledge and perspectives they can provide in other ways and include that information in the process.

- **Documenting and Communicating – Not Representing**

  To support this initiative and engage the entire workforce, electronic documentation and communication must be made accessible to all members of the organization. Open communication creates inclusion, avoiding the sense of "us versus them" that can characterize an effort in which one group of people develops a strategy in isolation from those called upon to enact it. We call this communication vehicle "the Playbook." One person needs to be responsible for its design, functionality and information flows. This coordinator, who is not a Team member, works closely with the Team to shape the content. Much of the information and communications between the Team and other organizational members will occur by way of the Playbook.[2]

  The Playbook serves as an electronic document that captures the process, the decisions made, the actions taken to implement the decisions and the results of those actions. It is a powerful method of keeping everyone in the organization informed and engaged from the beginning. Comments, suggestions and ideas are captured and the data mined so that nothing is lost. All entries become part of the electronic record of the initiative. We also suggest an actual or electronic wall[3] for sharing the reactions and comments coming from the organizational community.

  Team members in addition communicate with their colleagues through face-to-face conversation and via email and other media throughout the process. In this way, Team members personalize the initiative. They continually check in with their colleagues throughout to share information, test ideas and report on results. They communicate both horizontally with peer groups and vertically with their management and reports. They may speak at staff meetings, have one-on-ones with their reports and managers and meet with "communities of practice" [i.e., people with common interests such as engineers, lawyers accountants].

---

[2] The Playbook may have several forms and components, including a website on the intranet, blog, wiki, newsletters, surveys, etc.

[3] See the Repower America's site for an example: http://www.repoweramerica.org/wall/

Convening

Thus, the communication network is redundant, complex and overlapping. Employees may prefer certain forms of communication. They may miss some messages but pick up the information elsewhere in a different format.

While Team members are reporters and communicators, they are not representatives or negotiators of particular organizational units or points of view. They listen to responses and comments. However, it is important for them to realize that they are under no obligation to represent anyone else. If Team members feel that an important point needs to be added to the discussion, they will introduce it. However, they must be free to speak honestly for themselves. They are not speaking for their colleagues. They are simply representing themselves and expressing their own knowledge, views and perspectives. They have been selected as agents of organizational learning. Because the composition of the Team is inclusive, the Team itself encompasses a wide range of views. People will naturally reflect the perspectives of the functions and specialties they come from without acting in a formal representational capacity. Everyone in the organization has the opportunity to engage directly in the process through the Playbook. In this way, their ideas reliably and accurately reach the Team.

- **Modeling a Life-Sustaining Work Environment**

  The Team needs to understand that it is okay to fail, fail rapidly and fail often. Failure is essential to any creative process. By failing quickly, we learn quickly. Structural Dynamics provides the context to fail without the consequences of doing so in the "real world." In this setting, Team members feel safe to say what is on their minds, test out audacious thoughts, be a bit ridiculous and "throw ideas at the wall to see what sticks." Having the security to take creative risks allows the Team to innovate and experiment.

  Creating this space involves more than decor. Here are some factors that contribute to creating that sense of safety:
    - Roles and reporting relationships are set aside.
    - Everyone's view is equally validated.
    - Thinking and ideas are tested, not judged.
    - The physical surroundings are comfortable and casual.
    - There is plenty of room for moving around, posting ideas and sorting information.
    - When the Team is working in small groups, the groups can be in the same space, creating a buzz that enhances energy and excitement.
    - The space should have sound-absorbing materials in walls, floor and/or ceiling to reduce noise levels.

  If the documentation the Team generates – such as flip charts – cannot stay in a dedicated place between sessions, materials should be stored and redisplayed at the following session.

Thinking these matters through and having as much support as possible will be important to the creativity of Team members and the success of the initiative. The Team leader has to be particularly sensitive to the multitude of duties and priorities vying for Team members' attention. Job responsibilities can pull Team members away from this strategic work or, conversely, such responsibilities may suffer. The Team leader can support the members by developing agreements with them regarding such things as required resources, the frequency of sessions, research responsibilities and expectations for communicating with colleagues.

Approximately 25% of a Team member's time will be dedicated to this initiative as it proceeds through several phases. The actual time commitment will vary by the frequency of the meetings, the intensity of the effort and the amount of support provided. Developing and maintaining interactive electronic communication tools will facilitate the Team's work.

Convening

During the first Team meeting, members become acquainted with one another, learn more about their roles and responsibilities, and develop a better understanding of the Structural Dynamics process and the goals of this life-sustaining initiative. They also learn to think of the organization as a living system. They begin to realize the opportunities for personal development that can come from participation with this group of extraordinary people – in particular, the development of leadership skills. Specific agenda items for this meeting include:

- ☐ Articulating the Decision Issue to be addressed
- ☐ Selecting a time horizon for the inquiry
- ☐ Developing criteria for the Playbook
- ☐ Making agreements and establishing expectations

- **Articulating the Decision Issue**

    The Team must clearly grasp the necessity of being a life-sustaining system. Since the Industrial Revolution, organizations have thought of themselves as mechanistic structures. Some organizations are beginning to understand themselves as complex living systems. Knowing that the organization is similar in nature to the people who comprise it and the natural environment creates a powerful alignment. Without a life-sustaining point of view, organizations are at risk of exhausting their people and the planet's natural resources – both of which are essential to their own survival. This meeting should include a discussion of organizations as a living system and why it is crucial to be life-sustaining.

    Team members' first task is to clearly articulate the Decision Issue that will be the focus of their work together. The Decision Issue defines the scope of the initiative. A commitment to creating a life-sustaining organization is a broad objective. With such a vast scope, the Team runs of the risk of becoming unfocused. The group may have trouble knowing where to start, what needs to be included and what can be excluded. Members may conclude that, "Everything needs to be considered because everything is relevant."

    To make the initiative actionable, the Team needs to focus on a specific issue of fundamental importance to the organization. Decision Issues may vary widely. Each organization's situation [e.g., its industry, purpose, market position, mission, size, maturity, reputation, etc.] will shape the inquiry and determine the appropriate issue for the Team to address. Here is an example of a Decision Issue:

    > *What kind of environment, technologies, policies and practices do we need to provide for our people to succeed in an increasingly interconnected global economy?*

Convening

The outcomes of the process will inevitably touch on and affect a much broader range of issues than the specific one that the Team selects. A potent Decision Issue has a multitude of organizational ramifications.

The Decision Issue in the example above can have implications for...
- Debt and cash flow
- Formation of partnerships, mergers and acquisitions
- Relationships with partners and suppliers
- Co-location or dispersion of the workforce, etc.

Additionally, a strategic question such as this can prompt...
- Alliances with universities and research institutions to develop talent pipelines and enhance learning within the organization
- New facets of organizational culture and shared understandings
- More permeability between the organization and its external influences and resources

The purpose of narrowing the focus is simply to contain the inquiry so that it doesn't become amorphous or expand into an overwhelming challenge. The Team needs to be able to retain a focus on this issue throughout the process and articulate it crisply when communicating within the organization.

- **Selecting a Time Horizon**
  The time horizon is the period it will take for the results of changes made as a result of this initiative to fully play out and produce benefits for the organization. It is not the duration of the change process itself, but rather similar to the "payback" period of an investment. For example, while it might take several months to fit out a new office space, the life expectancy of that space, its "time horizon," might be 15 years or more. The nature of the Decision Issue will establish the time horizon. Although a 5-15-year perspective is a relatively common time horizon for life-sustaining organizational work, the specifics of the situation may require a longer or shorter view.

- **Establishing Criteria for the Playbook**
  The Playbook serves to both document the process and to provide two-way communication internally. In this session, the Team members specify the functionality of the Playbook in enough detail so that the Playbook coordinator can set up the intranet site and the group can be ready to start using it at the next meeting.

- **Making Agreements and Setting Expectations**
  The Team needs to discuss time commitments, spacing of sessions, expectations between sessions and norms of behavior. Several norms or "ground rules" are essential. These include:
  - ✓ Active listening to each other with full attention; not interrupting or mentally preparing a response while someone is talking.
  - ✓ Respect for the opinions of others even when they are upsetting or counterintuitive. Diverse opinions spark thinking and can be woven into alternative future scenarios.
  - ✓ Full participation at every step by each Team member.
  - ✓ Not taking up so much air time that others don't have a chance to participate.

Convening

These skills may require some development and practice to ensure that Team members have a consistent level of competency. The facilitator should review these points with Team members to enlist their agreement. They will undoubtedly suggest other rules that they would like to add to the list.

Although the Structural Dynamics process depends on a broad range of viewpoints, Team members should NOT be asked to state a position on the Decision Issue in this session. Putting a stake in the ground might lead Team members to feel defensive and cling firmly to their initial positions or feel the need to express their ideas aggressively. They may be reluctant to modify their views or seriously consider the views of others. Such a situation inevitably causes divisiveness and polarization.

For the life-sustaining initiative to succeed, Team members must understand from the start that the process belongs to everyone in the organization; it does not belong to the Team. The facilitator should coach Team members how to avoid any sense of exclusivity, cliquishness or special status. Inevitably, the process will entail its own lingo and jargon. If members need to use such terms when communicating outside the Team, they should take great care to explain the meaning and implications of these terms. The Team should always check to ensure that others are familiar with the content and flow of the process before blithely tossing around new phrases and concepts. It is the Team's responsibility to ensure that everyone in the organization has a chance to become familiar with the context in which the organization's operations can become life-sustaining, has the opportunity to contribute to the formation of the strategies and to decide how to best implement them for the benefit of whole.

With a strong sense of purpose, a compelling Decision Issue, a relevant time horizon, a sense of the Playbook format, clear expectations and agreed upon norms, the Team is ready to begin exploring the forces affecting the Decision Issue.

Exploring Facts

- **Role of Leaders: Futurist**
  - Your Experience
- **Scanning Events**
  - Safe Bets
  - Focus of Attention
  - Events, Variables, Trends
  - As Simple as Possible, But Not Simpler
- **Recognizing Patterns**
  - Complexity
  - Trends to Patterns
- **Fathoming Structure**
  - Patterns to Structure
  - Leverage Points
  - Interventions
- **The Heart of Structural Dynamics**
  - Pace of Change: Abrupt or Gradual
  - Response to Change: Reactive or Creative
  - Integrating the Dimensions
  - Scenario Archetypes
    - ☐ Maintaining the Status Quo
    - ☐ Exercising Discipline
    - ☐ Stepping into a New Reality
    - ☐ Experiencing Collapse
  - Scenario Game Board
  - Being a Player
- **Exploring Deeper**
- **Executive Briefing: Facts**

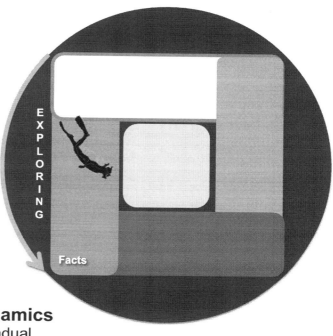

Exploring

 Role of Leaders: Futurist

> Nothing ought to be unexpected to us. Our minds should be sent forward in advance to meet all the problems, and we should consider, not what is wont to happen, but what can happen.
>
> *Lucius Annaeus Seneca, Roman philosopher*
> *3 BCE – 65 AD*

We can't know the future, but there are usually signals in the present indicating how the future could play out. For instance, astronomers have developed an array of devices to listen to the universe for clues about unusual occurrences, like the unlikely possibility of an asteroid striking Earth. Likewise, an organization's workforce is extremely skilled at listening and understanding, if they have a process that helps them make sense of what they observe and intuit. In a world in which we are constantly bombarded by a whirling, cacophony of information, the advantage goes to those who are attuned to the possible. When we explore the structure that drives past and present events, we learn that the same forces are shaping future events. We won't see the exact shape of the future, but we will understand how events in the present have causal connections to how the future will emerge.

Despite overwhelming evidence to the contrary, most of us implicitly assume a relatively modest pace of change. We live in a highly dynamic surround of factors that can be overwhelming. We frequently don't know how to influence events to favor the outcomes we want. As a result, we may simply rely on what has worked in the past. In doing so, we make ourselves vulnerable to being continuously surprised by events that seem obvious in hindsight and might have been anticipated with foresight. To effectively shape our future, we have to be just as agile and dynamic as our context.

Future-oriented leaders continuously scan the external environment for events and trends that may emerge as major forces in reshaping their organization's context. Rather than screen out information that doesn't fit their current worldview, they seek it. They read broadly outside of familiar domains, exploring social, economic, technical and political developments. They seek ideas from both conventional sources and those on the periphery of mainstream thinking. They enjoy conversations and interactions that expand their minds and hearts. Such leaders are "big-picture" thinkers.

Leaders who consider the driving forces shaping the future realize that these forces have causal relationships to one another. In doing so, they become whole systems thinkers. They are better able to anticipate changes that might affect an individual's full engagement with work, the ability of the organization to operate at maximum capacity and the way the organization contributes to the health of the natural environment. These are self-reinforcing attributes: a results-oriented organization that contributes to the natural environment will attract creative, talented people. These people produce results that galvanize positive action. By linking dynamic people with exciting products, future-focused leaders catalyze elegant solutions. They build robust living systems capable of sustaining life both inside and outside the organization.

> "I don't think anyone anticipated the breach of the levees.
>
> *George W. Bush*
> *Former U.S. President*
> *Referring to Hurricane Katrina, 2005*

Exploring

Futurists consider an organization's external environment in order to understand the forces shaping its context. Leaders, acting in their capacity as futurists, support and mentor the Team in its efforts to make sense of a broad range of disparate information that may pose a threat, present an opportunity, play to a strength or expose a weakness. They study factors and forces outside an organization's direct control but critical to its viability and success.

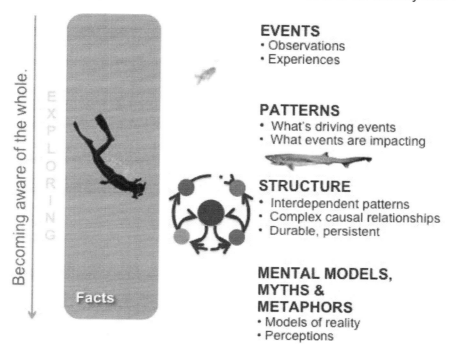

EVENTS
- Observations
- Experiences

PATTERNS
- What's driving events
- What events are impacting

STRUCTURE
- Interdependent patterns
- Complex causal relationships
- Durable, persistent

MENTAL MODELS, MYTHS & METAPHORS
- Models of reality
- Perceptions

Futurists look at seemingly disconnected **events** in the news, in society and in the workplace. Delving below the surface, they recognize **patterns** of events which repeat over time and they see that these patterns are interconnected in complex, dynamic **structures** which are influenced by the deeply entrenched **mental models, myths** and **metaphors** through which people view reality.

The dynamic structure of the natural environment and organization's external environment affects the workplace environment [e.g., work flow processes, work space, productivity and communication tools, performance pressures, staff interactions, power struggles, aspirations, policies, culture, norms...].

Futurists are comfortable dealing with the creative tension created by complexity, ambiguity and contradiction of divergent factors. These forces and elements form a complex weave of possibilities and challenges. People, places and countervailing plot lines are the threads they weave together to create compelling alternative futures. Seeking "quick-fix" solutions over-simplifies the richness of reality, thereby glossing over both the challenges and the opportunities.

Exploring

Futurists use a whole systems thinking perspective to lead others to an understanding of how events are connected through interacting patterns driven by dynamic structural forces. Grasping these concepts provides a sense of how these forces may drive events in the future.

As people within the organization develop a shared understanding of the organization's strategic situation, they act more in concert with one another. Understanding the organization's current reality creates a bond. When people learn together, their organization becomes more cohesive, whole and life-sustaining.

- **Your Experience**

    The following self-assessment tool can help the reader gauge the degree to which you are presently inclined to think and act as a futurist. Futurists' skills and thought patterns are eclectic, varying widely from person to person. Nevertheless, this inventory may point to strengths that you already have and indicate areas that you may want to emphasize more in your thinking and leadership. Rate your competence in each of these domains using a numeric scale.

    Futurist Self-Assessment Tool

    - ☐ I regularly and systemically scan and analyze information on a wide range of topics, including those that are unfamiliar.
    - ☐ I expose myself to unique, perspective-enhancing experiences and people.
    - ☐ I am willing and eager to look past what is already known to what might be possible.
    - ☐ I constantly play out how the future might unfold by thinking, writing and talking about the subject and soliciting a variety of views.
    - ☐ I am at ease considering alternative futures; I am not stuck on one particular version of the future.
    - ☐ I am able to describe and critique my own mental models of how the world works.
    - ☐ I live as if very different futures could unfold from the facts of the present.

    You may find it useful to make note of examples of times you've displayed the skills of a futurist and when you may have missed an opportunity. Look for ways to practice particular competencies that you may not have focused on previously.

Exploring

💡 Scanning Events

Events that are not within the control of the organization nevertheless shape its environment. We read about these events in newspaper headlines, hear about them on the evening news and experience them in our daily lives. Events are going on around us constantly – incidents, issues, problems. They often seem disconnected and discontinuous. This apparent randomness can lead to feelings of helplessness or apathy. We believe others are, or should be, doing something about the larger picture because it is clearly out of our control.

Politicians and media outlets often select and exaggerate events to make a point, defend a position or shape the thinking of others. This selectivity, required by the sound-bite format, disguises the wholeness of reality, presenting it as discrete, separate parts. For example, the evening news reports on one climate disaster after another and rarely, if ever, puts the events into a systems perspective, explaining the forces driving unusual climate patterns. In reality, these stories are not isolated from one another any more than waves are separate from the forces that cause them to occur. A pattern of waves is driven by forces above and below the surface [e.g., lunar position, storms, currents, earthquakes, etc.]. Events in the news are also driven by a number of interacting forces. By exploring below the surface of events, we can understand the structure of the forces causing them to occur.

The commonly used phrase, "connecting the dots," refers to grasping the relationships between events. The Structural Dynamics process starts by "collecting the dots" that will later be connected. An organization does so by scanning for events that might be relevant to its ability to become life-sustaining. By exploring for such events, people develop a capacity to think broadly and be open to new information from wherever it may come.

So that we don't lose focus amid the vast array of events that we might consider, we use a "framing question" that links the exploration to a Decision Issue we want to consider and bounds it within a specific time horizon. If, for example, we were working with the Decision Issue:

What kind of physical, social and cultural work environment do we need to provide for our people to operate at their full potential?

The framing question would include the Decision Issue and the timeframe:

What external events, if they happen in the next 10 years, would have a significant impact on the kind of physical, social and cultural work environment we need to provide for our people to operate at their full potential?

The framing question focuses attention on the most relevant events.

Exploring

When scanning for events, we recommend using the categories of Societal, Technical, Economic, Educational and Political: the first letters of these words spell STEEP. We find this approach useful, particularly when we add Environmental and Aesthetic to make it STEEEPA. Keeping these categories in mind helps to broaden the types of events explored:

- ✓ <u>Societal</u> events relate to social phenomena and activities such as demographics, expectations, preferences, cultural norms, beliefs, attitudes, fads, etc.
- ✓ <u>Technical</u> events, include biotech, nanotech, information technology, etc., related to technological advancements or concerns.
- ✓ <u>Economic</u> events relate to the aggregation, dissipation and flow of money, finance, industry, labor, etc.
- ✓ <u>Educational</u> events have to do with educational achievement, the process of learning, learning methods and the functioning of educational institutions.
- ✓ <u>Environmental</u> events relate to the condition or well-being of nature including flora, fauna, atmosphere, land, water, weather, climates, the balance of natural ecosystems and extreme weather events such as earthquakes, hurricanes, tsunamis and tornadoes.
- ✓ <u>Political</u> events relate to ideologies and powers shaping the context of an organization; the regulatory and legal structure governing the operations of local, regional, national and international ruling bodies; global trading blocks, unions, agreements and pacts; and the dynamics of power between competing or allied stakeholders.
- ✓ <u>Aesthetic</u> events involve the performing, graphic and fine arts, including styles; craftsmanship; the use of color, form, texture and light to create an experience; a cohesive or discordant look and feel of space, technology and personal expression; and activity or movement in the fields of the arts and design.

Examples of Events

- ▸ Availability of universal health insurance in the U.S. allows workers to change jobs or launch a small business without losing access to affordable health coverage.
- ▸ Video-teleconferencing technology becomes inexpensive, ubiquitous and reliable.
- ▸ California prohibits relocating back into areas evacuated twice in five years due to natural disasters: 10,000 households and businesses in Los Angeles are displaced following the mudslides of 2015.
- ▸ China launches satellites intended to supply 10% of its power requirements through solar energy.
- ▸ Jon Stewart of Comedy Central becomes CEO of the New York Times Company.
- ▸ By 2020, half of the doctors, engineers and other professionals over age 65 around the world continue to work well into their 80s and 90s, at least part time.
- ▸ Deteriorating infrastructure in many parts of the developed world causes major disruption in transportation, communication, utility distribution and sanitation systems.

Exploring

A Team brainstorming session might generate hundreds of events. As we have seen, the process deliberately encourages a range of views. Many different, sometimes conflicting, perceptions and ideas in response to the framing question are to be expected. Contradiction is desirable; indeed, the process thrives on it. In every group with which we have worked, thoughtful, intelligent and reasonable people anticipate very different future conditions. Such divergence illustrates the uncertainty of the future. Something that seems nearly assured to some members of the group [e.g., advances in artificial intelligence that would make computers better decision makers than humans] might seem like science fiction to others. These disparate views are the basis of distinct scenarios of the future.

Although there is value in encouraging the Team to read broadly and collect ideas about anything that might affect the Decision Issue they identified in the first session, it is best to generate a list of relevant events together as a group. If individuals come into the conversation with a list that they want the group to consider, people's positions may harden. When people look at an issue freshly together, they see themselves as a community, building a joint understanding of the situation. Individuals with firm beliefs will start to hear alternative views. Having people work together can also increase the yield of ideas; events will be generated at a faster rate, one building on another.

Notwithstanding the value of thinking collectively, we have found value in asking participants to reflect individually on events for a few minutes before the group brainstorming session. This moment of introspection ensures that everyone has centered themselves in the process and formulated something that the group will take into account. This reflection period should be no more than a few minutes.

The facilitators do not need to qualify events while brainstorming except to test that they respond to the framing question based on the Decision Issue. They will want to be aware of the factors that can influence which events come most easily to mind:

→ Dramatic current events in the news or in the organization capture our attention and color our view of the future.
  ▸ In times of crisis, people tend to generate events that are fraught with anxiety and pessimism. For example, in the midst of a financial crisis, we will come up with events related to economic downturn.
  ▸ In times of prosperity, events tend to suggest that blue skies will go on forever – we have a hard time imagining that growth will ever stop.

→ Events expected to occur soon loom larger in our minds than events that might occur in the remote future.

→ Events that threaten or entail costs or risks come to mind more readily than those that might benefit the issue under consideration [e.g., the expense of repairing a deteriorating physical infrastructure gets more attention than the possibility of enlivening of a depressed neighborhood].

→ Media coverage can focus our awareness on relatively minor events and distract our attention from events that are more important. For example, a nuclear peace accord gains less time on the evening news than the latest personal scandal in Congress.

Skillful facilitation can guide the attention of the Team in ways that mitigates these potential imbalances. For example, early in the generation of events, facilitators can probe for occurrences that might occur toward the end of the time horizon that has been established.

Exploring

This brings far-off possibilities closer into view and moves the brainstorming session beyond those events that are top-of-mind [Gertner 2009].

- **Safe Bets: Certainties**

    Is anything certain to happen? Certainties are those events affecting the Decision Issue that *will* happen no matter how the future evolves. For example, we can calculate the number of graduates in various fields based on college enrollments today. Demographics often provide a source of inevitabilities.

    One organization we worked with called certainties *freight trains* because "they are coming down the tracks whether you want them to or not." Their certainties included a major change in their traditional business model [how they charged for their products] based on technical and competitive forces. Even if they could not predict how or when the change would occur, they knew that such change was inevitable within the timeframe they had identified.

    Certainties can be thought of as future "facts," but it is difficult to see inevitabilities that we are not currently experiencing. Here are some events that strategic thinkers, including us, believe *might* well take place in the next 25 years:
    - Dramatically increased life expectancy
    - Healthier old age
    - Multiple generations in the workforce
    - Increased frequency of catastrophic climate events
    - Economically viable, renewable energy sources
    - More integration of work and leisure
    - More contracting and freelancing
    - Emphasis on capability and performance over hours, location and appearance
    - Boundaryless collaboration
    - Dramatic resource constraints

    Are they certainties? No. Certainties are rare. Just because a lot of people think the same way, doesn't make it certain. A survey of predictions made by professional futurists when the present was the future, confirms the difficulty of identifying certainties.

    And, we must be careful not to mistake events and conditions to which we have grown accustomed for certainties. For example, U.S. housing prices rose steadily since the 1950s and we came to believe that this trend was inevitable – until 2008. Like many things we have come to know through experience, we need to reexamine our assumptions about what is certain.

    In order for an event to be deemed a certainty, all Team members must agree; any disagreement indicates a lack of certainty. If any certainties are identified, they will, by their very nature, occur consistently in any future that the team considers; after all, we have said that these things *will* occur no matter what else happens.

    Realizing that something is bound to happen provides strategic insight. Knowing a few facts about the future puts an organization on more solid ground in its strategic thinking, planning and action.

Exploring

- **Focus of Attention: Critical Uncertainties**

The vast majority of events are uncertain. How can we determine which ones deserve our close attention? How do we know which ones will change our lives, our work and our organizations? In M.C. Escher's "Day and Night," our perception of the scene shifts dramatically depending on how we view the situation. Humans may be on the brink of breakthroughs in science, medicine and technology such as robotics, genomics, nanotechnology, etc. [Canton 2006]. *And,* we may simultaneously be on the brink of catastrophic interrelated economic, climatic and conflict crises.

To focus our attention where it is most needed, the next step in Structural Dynamics is to sort through the uncertainties to identify those that are highly critical to the Decision Issue and are also highly uncertain. Critical uncertainties have a huge potential impact on the Decision Issue, but they could occur in a number of different ways. Each organization arrives at its own set of critical uncertainties. The critical uncertainties of the utility industry are likely to be quite different from those faced by a professional nursing association, for example. And there will also be some common factors. Both utilities and medicine are affected by the level of government regulation, for example.

Consider the types of benefits employees will require from organizations competing for their talent. The current trend toward fewer employer-paid benefits could change. If so, what will an organization need to provide? Health benefits for aging parents? Satellite offices in assisted-living facilities?

Critical uncertainties form the basis for strategy development. Because of their criticality combined with a high degree of uncertainty, these factors tend to generate disagreement and confusion. A critical uncertainty is like a fulcrum; the organization that can identify key leverage points is able to influence events in the direction of its preferred future.

Exploring

- **Events, Variables, Trends**
  When we look more closely at seemingly isolated events, we often find that they have underlying *variables* that change over time. Prices rise or fall. Sales grow or decline. Supply of skilled labor is tight or it is plentiful. We can graph the changes in these variables as trend lines.

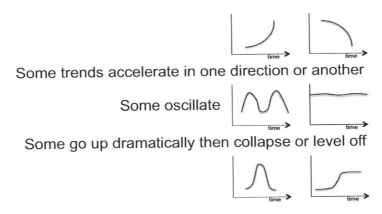

Some trends accelerate in one direction or another

Some oscillate

Some go up dramatically then collapse or level off

Below is an example of an impressionistic trend line. It indicates the number of women in traditionally male roles [e.g., physicians] as a percentage of the total number of people in those roles over the past 10 years.

EXAMPLE: Trending an Event

- Future Event:
  [time horizon = 10 years]
  In 2017, for the first time in history, MIT* has more tenured women on the faculty than men.

- Underlying Variable:
  Women in traditionally male professions

- Historic Trend:
  [looking back 10 years]

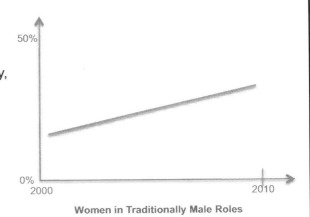

Women in Traditionally Male Roles

* The Massachusetts Institute of Technology in Cambridge, Massachusetts, was founded in 1861. Women have been admitted since 1871; however, before the 1964 opening of MIT's first women's dormitory, only a handful of women attended. A woman became a tenured professor for the first time in 1963. In the fall of 2008, 45% of undergraduates were women and 31% of the graduate students.

If it provides useful information, a variable can be graphed precisely with researched data. Often, it is sufficient to simply sketch the group's impression of how a variable has been trending in the past equal to the time horizon identified by the Team.

We cannot assume that trends will continue in the same direction as they have in the recent past. In this process, we look backwards to see how a variable has been trending, not necessarily as an indication of where it will go in the future. We will see that many forces acting on the variable will influence the trend going forward.

- **As Simple as Possible, But Not Simpler**

  If an event has several underlying variables, the group can aggregate those factors or articulate them individually. The appropriate level of specificity depends on the Decision Issue. For example, the variable "Women in Traditionally Male roles" encompasses women:

  ▸ professionals – lawyers, architects, physicians, etc.
  ▸ in the trades – plumbing, construction, etc.
  ▸ scientists and technologists
  ▸ academics – deans, professors, presidents, etc.
  ▸ executives – CEO, CTO, etc.
  ▸ board members
  ▸ holding elected office – senators, presidents, governors, mayors, etc.

The crucial test is to be sure that any additional detail improves the value of the analysis more than it complicates it. Every new factor increases the complexity of the analysis. Include only those variables that provide more clarity than confusion. The analysis should be rich in the details that add value but free of extraneous subplots, gratuitous images, random thoughts and side issues. This idea of being "as simple as possible but no simpler" is known as Occam's Razor, attributed to scholastic philosopher William of Occam.

> The explanation of any phenomenon should make as few assumptions as possible, eliminating those that make no difference in the observable predictions of the explanatory hypothesis or theory.

William of Occam
c.1288 – c.1348

Exploring

🌀 Recognizing Patterns

An awareness of variables and trends increases our ability to recognize patterns. A pattern is a potent field of connections among related variables. It manifests the dynamic impact of the variables on one another.

Buckminster Fuller referred to his hand as having "pattern integrity," which he defined as "the universe's capability to create hands." [Kenner 1973]

A pattern is in integrity with itself. It is not acting from any special sense of purpose. It is conforming to the laws that connect the variables within it. It is the mathematical abstraction that accompanies form.

The wave is not the water.
The water told you about the wave going by.
But the wave has a patterned integrity of its own
— absolutely weightless.

Buckminster "Bucky" Fuller
1895 – 1983

A set of interrelated forces comprising a pattern always manifests in a similar fashion. The relationships between the variables are durable – they have existed in the past, persist in the present and can be expected to endure into the future. These causal relationships recur over and over with infinite variations in response to particular conditions. No two hands or two waves are exactly the same, but each follows a given pattern; for instance, a child's hand grows into an adult hand.

In *A Pattern Language* [1977], Christopher Alexander identified hundreds of patterns that people have used throughout history and across cultures to create their built environments in ways that are life affirming.[4] Although "all of the ugliest, deadliest places in the world are made from patterns as well," Alexander concentrates on patterns that result in soul-satisfying places. Anyone can apply the patterns he identifies to create life-sustaining, rather than deadening, environments. By establishing persistent life-sustaining patterns, organizations can be as nourishing to the soul as a wave, a crystal clear day in winter or a rocky coastline.

The patterns repeat themselves because, under a given set of circumstances, there are always certain fields of relationships which are most nearly well adapted to the forces which exist. The shape of the wave is generated by the dynamics of the water, and it repeats itself wherever these dynamics occur.

Christopher Alexander, 1979

---

[4] For example, Alexander has noted that homes throughout the world have a privacy gradient: living rooms meant for visitors are placed toward the front of the house and the degree of privacy increases as one goes deeper into the home.

Exploring

- **Complexity**
A mechanistic perspective assumes that every cause produces a single effect, occurring closely in time and space; an input results in an output. If the world were mechanical, the future would be predictable: we would have a crystal ball into the future. All that would be necessary to create a life-sustaining organization would be knowable in advance with precision and certainty.

Of course, it's not that simple. Complexity and chaos theories reveal a reality that is much messier than that viewed through a mechanistic lens. Complex dynamic systems are those in which nothing ever happens in quite the same way twice, and yet enough happens in a tidy enough way to preclude complete and utter havoc. In chaotic systems, there are multiple causes and multiple effects. The influences are profuse and diffuse. The elements of the system are in constant motion. Awareness of the dynamic nature of causal relationships is a powerful way of seeing and acting in our increasingly complex and always chaotic world.

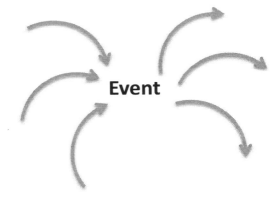

All things come into being only as a result of the interaction between causes and conditions. They don't just arise from nowhere fully formed. There is mutual dependence between parts and the whole. Without a whole there can be no parts; without parts, there can be no whole. Anything that exists does so only within the total network of everything that has a possible or potential relationship to it. No phenomenon exists with an intrinsic or independent identity, and the world is made up of a network of complex interrelations. We cannot speak of an independent reality without speaking of its range of relations with an environment and other phenomena, including language, concepts, and other conventions.

The Dalai Lama

Exploring

Example of a Pattern:
The following graphic shows a pattern of variables that have the effect of reducing the amount of real estate required by an organization.

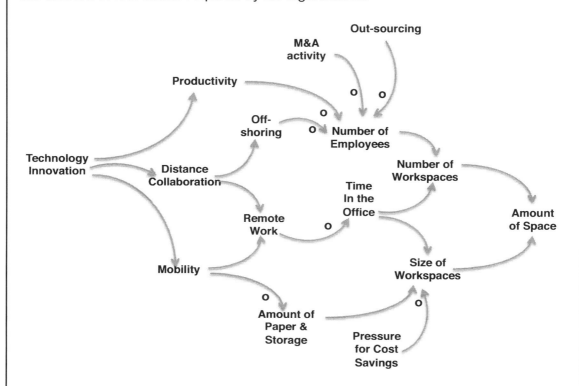

Note: The "o" indicates that the linked variables move in opposite directions. The linked variables without an "o" move in the same direction.

Stepping through this pattern starting from the left side:
- Technology Innovation
  - improves Productivity,
  - allows more Distance Collaboration and
  - increases Mobility [ability to work anytime/anywhere inside or outside the office]
- Distance Collaboration allows an organization to do more Off-shoring
- Off-shoring, Productivity gains, M&A activity and Out-sourcing lead to a reduction in the Number of Employees
- Distance Collaboration and Mobility allow more Remote Work
- Remote Work reduces Time in the Office
- Fewer Number of Employees and less Time in the Office reduces the required Number of Workspaces
- Mobility leads to a reduction in the amount of Paper & Storage [less to carry]
- Reduced amount of Paper & Storage along with less Time in the Office and Pressure for Cost Savings reduces the Size of Workspaces
- Fewer Number of Workspaces coupled with smaller Size of Workspaces leads to a reduction in the Amount of Space required by an organization

Indeed, we are seeing a decrease in the amount of space occupied by organizations in relation to the number of people they employ. In 2006, the U.S. Government Services Administration, Office of Real Property Management, reported that the combined effects of telework and hoteling had reduced its space requirements by 8% through decreased numbers of workstations.

On the other hand, certain forces might drive a need to *increase* the amount of space. These include:
- the necessity for teams and groups to meet face-to-face to develop trust
- the benefits of being co-located with colleagues to share information and learn from each other on an ad hoc basis
- competition for key talent may necessitate dedicated space for employees

> Research shows that if improved workplace strategies were applied across the board to the entire Federal office space portfolio [comprised of 725 million square feet], the potential savings to the government could be as much as $8.4 billion annually. Applying this same potential formula for savings to all Federal space types [3.4 billion square feet] could ostensibly produce savings near $39 billion annually.
>
> *GSA Office of Governmentwide Policy*
> *Innovative Workplaces: Benefits and Best*

Other forces, too, would exacerbate the trend toward a decrease in space required, such as an international economic downturn or an organizational revenue decline. Any organization doing this type of analysis would have to be sure that it is not overlooking critical factors.

The cost savings in real estate could be channeled into other areas [e.g., collaboration tools, home office stipends, salaries, benefits] as the organization determines the best ways to create and maintain life-sustaining qualities.

- **Trends to Patterns**

    Let's go back to the variable, "Women in Traditionally Male Roles" in the developed world. We might determine that it is related to:
    - Women's Rights
    - Women's Education Level
    - Pay Equity

    Here is an impressionistic view of the trend lines for these variables:

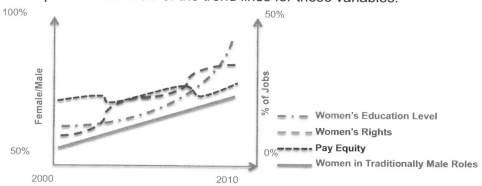

Exploring

The causal [cause and effect] connections between these variables can be shown as in this loop diagram…

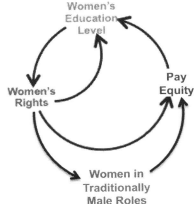

This is a set of nesting, reinforcing processes. A rise in Women's Education Level leads to more Women's Rights, leading to more Women in Traditionally Male Roles. An increase in Women's Rights and more Women in Traditionally Male Roles leads to more Pay Equity. Women's Rights and Pay Equity cause a rise in Women's Education Level.

These dynamics are reinforcing because all the variables move in the same direction – when one increases, the others increase; when one declines, the others decline. There may be significant delays in the system; it can take months, years or decades to see how a change in any one of these variables affects the other variables.

🌑 Fathoming Structure

A system of interacting patterns forms a structure. Even simple patterns can interact to form intricate structures. Complex systems can vary in many ways and create an array of possible outcomes. Sometimes, unnoticed or unseen structures dominate our thoughts and actions.

When we arrive at the structural level, we discover the essence of living systems. This "Aha!" moment is the basis for deep awareness. Once we understand that the behavior of a system is not random, we can determine what is driving it. Understanding how a structure behaves provides insight into which events might surface above the water line.[5] Such insight allows us to better anticipate the future. By grasping the structure, we understand how our actions impact the dynamics of the situation.

> In the sci-fi film, *The Matrix,* the protagonist, Neo, is being pursued by a band of "agents" intent on killing him. After quite a bit of training and questing, Neo has a flash of insight that enables him to see the structure of the situation…where the agents come from, why they are shooting, why they run after him and why he runs from them. These events are the result of an underlying structure of systemic forces. With this insight, Neo has an understanding of how best to intervene to change the impact of the dynamic forces in play. He turns to face his pursuers, calmly holds out his hand and the agents' bullets fall to the ground, devoid of their momentum. Their ability to terrorize him depended on his complicity.

---

[5] Refer to the graphic on page 39.

- **Patterns to Structure**

    It is not a good idea to simply project historic trend lines into the future. Trends eventually meet with one or more countervailing forces that will change the course of the trajectory, slowing it, pausing it or turning it around.

    Going back to our previous example, many of us tend to assume that women's rights will continue to grow. Women are serving at virtually every level of government, e.g., German Chancellor Angela Merkel, U.S. Secretary of State Hillary Clinton and Argentine President Cristina Fernández de Kirchner [left]. Academic achievement by women is outpacing that of men in a number of countries. Female athletes are setting strength and endurance records and excelling at sports once deemed off limits. The list of achievements is very long. It seems that the genie is out of the bottle and there's no going back!

    However, that may not be the case. In many cultures and sects, a rise in women's rights is perceived as a threat to the status quo and leads to an increase in religious mandates, formal edicts or informal norms, often powerfully supported by religious beliefs. These forces can lead to a decline in Women's Rights. The imposition of Patriarchal Traditions is, therefore, a limiting factor on the unfettered expansion of Women's Rights.

    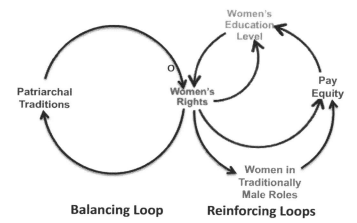

    Balancing Loop   **Reinforcing Loops**

    Because the variable Patriarchal Traditions moves in the opposite direction from Women's Rights [shown by the "o" for opposite], a "balancing loop" is created that tends to keep the growth of Women's Rights in check. Over time, periods in which Women's Rights are strong alternate with those in which Patriarchal Traditions dominate, creating an oscillating equilibrium between these two forces. The period and magnitude of the oscillation depends on the relative strength of the various forces.

    The variable Patriarchal Traditions gives rise to events such as the following:
    ▸ In India, a "ladies' train" created to allow women to commute without harrassment is regularly boarded by men who berate them.
    ▸ In Iran, Guinea and many other societies, men in powerful government positions frequently lace their ideology with religious doctrine that specifies a subservient position for women in society and the family order.

Exploring

- In Jeruselem, some Orthodox Jews yell and throw things at women who are praying at the Wailing Wall.
- The Southern Baptist Convention in the U.S. establishes the submission of wives to to their husbands as a formal doctrine.
- Catholicism remains adamant that women cannot be priests or use contraceptives.
- Multiple studies demonstrate that the preference for male children is virtually universal.
- *The Economist* estimates that in China at least 100 million girls have been killed in a "gendercide."

Seen from this perspective, the expansion of women's rights appears far from guaranteed. Could it be that ten years from now, women might face a much more rigid set of conditions than they do at present? It is hard for most educated Westerners to imagine that we may have seen the zenith of the power of women in modern times. The powerful dynamic tension of forces in this structure can and will play out in many ways – sometimes expanding and other times contracting women's rights.

- **Leverage Points**

Living systems are in constant motion; they require constant inputs of energy to maintain order and to grow. They tend to stagnate and die when they reach equilibrium. They freeze, shatter and fade when there is no new energy to keep them alive. Attempting to bring a complex organization under centralized control is like trying to herd cats. Because an organization is a complex adaptive system, its leaders can guide and shape it only when they understand the dynamics of its structure. With knowledge of how the forces in the organizational system interact, leaders can identify the most effective means of action. Intervening by increasing or decreasing the pressure on key variables will affect the entire system. Even a small amount of pressure exerted at these critical points can have a powerful impact by amplifying into larger effects throughout the system. Conversely, enormous effort exerted in the wrong places can be totally ineffective. By knowing where and when to intervene, leaders can effect change and move the organization in the direction of becoming life-sustaining.

The structure of a system and its information flows determine behavior ... growth, stability, decay, success and failure. Internal feedback loop relationships cause a system to change through time. Understanding why a system behaves as it does permits redesign of structure and policies to improve behavior.

Jay Forrester, 2009

- **Interventions**

When looking for leverage points, keep in mind that there are two types of forces [or connections or information flows] in every system. One is a *natural connection* that links clouds to rain, A to B, yawning to sleeping, etc. These connections are based on the laws of nature and are difficult or impossible to change. The other type of connection is based on *choice.* These choices or decisions may be based on habit, preference and belief, or they may be based on information, knowledge and wisdom. For example, if a decline in sales leads to a decrease in revenue, executives may decide to reduce the number of employees to lower expense and boost revenue. The unintended consequence of this action is that the company may fall into a downward spiral if having fewer employees results in further declines in sales and revenue. If the executives instead make the difficult choice to invest scarce resources in product development or improved information systems, the result might be different. These choices could lead to an increase in sales and revenue over time. When we want to change the way things are going, interventions are much more likely to be effective if we target a choice point rather than a natural connection.

Structural Dynamics has many parallels to the *Theory U* [Scharmer 2009]. Visualize going down the left-hand side of a "U." Scharmer describes this phase of Theory U as "Observe, Observe, Observe," as we open ourselves to new possibilities and new ways of seeing. The bottom of the "U" represents "Retreat and Reflect, allowing inner knowing to emerge." Ascending the right-hand side of the "U" brings the new knowledge and insight into the world; the participant is ready to "Act in an Instant." In the Exploring phase of Structural Dynamics, we have been observing, observing, observing as we descend from the surface level of events to patterns and then to structure. We are learning and deepening our understanding of the people that comprise the organization, the system that is the organization and the larger system within which the organization operates. In the Discovery phase, we retreat to the future and allow inner knowing and elegant solutions to emerge. In the Embodying phase of Structural Dynamics, the organization's new knowledge and insight enables it to act with confidence "in an instant."

We are now going to present a structure that models four distinct scenarios of future possibilities. This set of scenarios is archetypal; it covers a broad range of possibilities and represents the way that people have viewed the future throughout history and across cultures. We will then go to a deeper level, exploring mental models, myths and metaphors – emotional and cultural givens that undergird the reality we perceive. These ways of thinking, which we all have, can thwart our intentions to be impartial observers and to keep our minds open to future possibilities.

Exploring

The Heart of Structural Dynamics: The Scenario Game Board

By looking at events, variables, trends and patterns, we are developing a sense of how things are at present in the organization's environment. How do we take that knowledge and project it into the future? Like a branch of a tree, the future can grow in any number of directions. We use scenarios to consider a broad range of future possibilities. A framework for the scenarios consists of two dimensions: one based on the laws of nature and the other on human choice.

- **Pace of Change: Abrupt or Gradual**
  In a living system, small, gradual changes are common. Infrequently, a massive, abrupt shift will take place. The forces that cause earthquakes, for example, gradually build up as tension in the tectonic plates and then Boom! This phenomenon is governed by laws of nature. Theoretical physicist Per Bak calls this "self-organized criticality" and provides a fascinating illustration using a pile of sand:

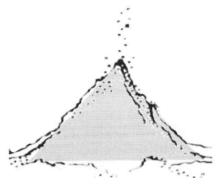

- When you first start building a sand pile on a tabletop, the system is weakly interactive. A sand grain drizzled from above onto a randomly chosen location on the tabletop has little effect on sand grains at other locations.
- However, as you keep dribbling sand grains from above onto randomly chosen tabletop locations, eventually the sand pile at one or more locations reaches a "critical state" where the pile cannot grow any higher without a breakdown of the pile occurring.
- These breakdowns can be of various sizes depending on the exact configuration of the sand pile at the time the breakdown occurs.
- Bak refers to these critical states as states of self-organized criticality [SOC], i.e., states in which the system has self-organized to a point where it is just barely stable. What does it mean to say that "breakdowns of all different sizes" can happen at an SOC state?
- The dribbling of one more grain of sand onto a location in an SOC state can result in an "avalanche" or "sand slide," i.e., a cascade of sand down the edges of the sand pile and [possibly] off the edge of the table.
- The size of this avalanche can range from one grain to catastrophic collapses involving large portions of the sand pile.
- Bak conjectured that the size distribution of these avalanches obeys a "Power Law" over any specified period of time.
- That is, he conjectured that the average frequency of a given size of avalanche is inversely proportional to some power of its size, implying that big avalanches are rare and small avalanches are frequent.

"Introduction to Self-Organized Criticality & Earthquakes"
Nathan Winslow, Department of Geological Sciences, University of Michigan

Exploring

There is no way to predict the effect of a particular grain of sand – whether it will just make the pile a little bigger or cause the entire structure to change. A self-organized criticality can be thought of as a tipping point [Gladwell 2002], a chain reaction or the straw that broke the camel's back. We understand the application of self-organized criticality to natural events such as avalanches, hurricanes and volcanoes. These concepts of gradual and abrupt change also apply to social, political, environmental and economic processes.

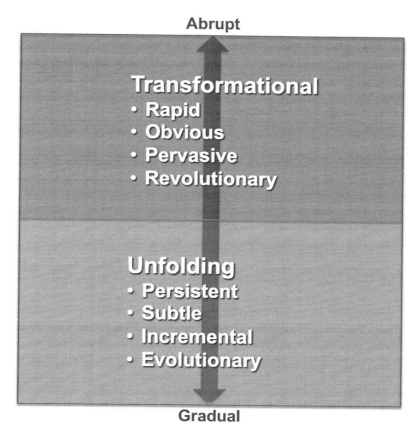

Abrupt changes can occur on the world stage:
- a slow decline leading to the sudden collapse of empires
- a gradual build-up of international tensions that erupt into war
- incremental changes toward a higher level of risk taking that result in global financial crisis
- experimentation with alternative energy sources that culminates in breakthroughs in science and technology [e.g. algae power, solar satellites, photovoltaics, biofuels, etc.]

Abrupt change can occur in an organization's environment:
- game-changing competitors enter the market
- tastes, fads and fashion affect demand for products/services
- regulation, taxation and trade policies impact business models

Exploring

And, abrupt change can occur in the workplace environment:
- loss of key staff members to death, accident or competitors
- acts of violence or natural disasters that disrupt operations
- hostile takeovers

Abrupt changes have a transformational effect. Once an abrupt change has been triggered, it cannot be stopped; there is no turning back. Once in motion, these forces are irreversible.

The effects of incremental change are apt to go unnoticed. Even when the speed of change is as glacially slow as an ice age, something is going on. And, although it is not always the case, these incremental changes can have an enormous impact. Like the metaphoric frog that boils to death because it fails to notice that the water temperature is gradually rising,[6] the impact can be as far reaching as transformational change.

Unfolding events at the global scale include:
- levels of carbon dioxide aggregating in the atmosphere
- nuclear proliferation
- power shifts in international blocks

Gradual change in an organization's environment might include:
- maturity of the industry
- obsolescence of products and services
- availability or scarcity of key skills and talent

Gradual change in the workplace environment might include:
- evolution of management style – hierarchical vs. networked
- the balance of face-to-face vs. remote interaction and communication
- the use of collaborative space versus private space

Change occurs gradually, in a smooth linear progression – unless it doesn't. Both abrupt and gradual change can have a great deal of uncertainty and indeterminacy associated with them. You can be certain that a transformative change will occur, for example, but you don't know when. You can be sure that there is a very small increase or decrease of some phenomenon every time you check, but you might not know if it's ever going to be of great importance.

---

[6] An apparently apocryphal story holds that a frog tossed into boiling water will immediately leap out. A frog sitting in cool water that is gradually heated will stay put while being thoroughly cooked. Many organizational failures and social breakdowns can be understood in terms of the boiled frog syndrome. See Gore's [2006] *An Inconvenient Truth* for a potent example.

- **Response to Change: Reactive or Creative**
  Another dimension, orthogonal to the pace of change, is our response to change – a "human choice." Such "choices" are emotional, psychological and often so deeply ingrained that we act in a seemingly automatic fashion without conscious thought or awareness – without even realizing that we have a choice. Our response might be dominated by fear or hope:

  - In a state of fear, an individual, organization or society views a change with anxiety. Emotions take over, sometimes accompanied by panic. We resort to fight or flight. We want to protect ourselves and those we care about. James Surowiecki, author of *The Wisdom of Crowds*, says, "Both history and theory suggest that tough economic times make people less interested in sharing burdens, not more" and "It is hard to pass reform programs...when voters are trying desperately to protect what they already have" [Surowiecki 2010].
  - If we experience change or potential change as an opportunity, a pathway to an exciting adventure and a new way to grow, our response is hopeful, proactive and creative. Our sense of community, the feeling that "we are all in this together," is enhanced.

  Both fear and hope are natural reactions. In any situation involving change, both forces are likely to be present – often in a single individual, group or organization.

Exploring

- **Integrating the Dimensions**
  Overlaying these dimensions, we get a two-by-two grid with four quadrants:

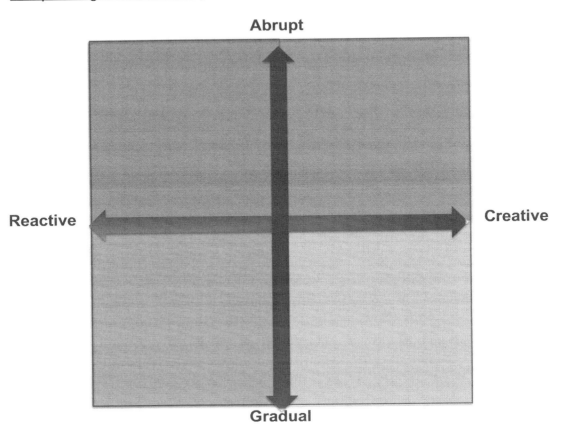

A distinct scenario, a story of possible future conditions, can be developed within each of these quadrants based on these dimensions. The specific content of each of the scenarios will, of course, be determined by the issues and conditions facing each organization and by the information, knowledge and wisdom gained by thinking together about them. Integrating "natural connections" and "human choices" creates a framework that lays out distinct future possibilities. Because these options cover real possibilities, we can start to appreciate why people hold such dramatically different visions of the future.

While this two-by-two matrix resembles those used in other scenario processes, there are two fundamental differences:
- Structural Dynamics always uses the same two dimensions [*pace of change* and *human response*] because we have found these to reliably provide a framework for archetypal scenarios that cover the images of the future that people of many cultures have used throughout time.

Exploring

> The resulting archetypal scenarios are based on an analysis of causal relationships and not simply imagination [see Scenario Game Board later in this chapter]

- **Scenario Archetypes**

    An infinite number of plausible future conditions exist. The two-by-two matrix points us toward a set of divergent scenarios that encompass many of those future possibilities. An organization can use this framework to order the array of possible futures into a set of images that enable it to plan for the future.

    Many of us have had the experience of a situation, point of view, art object, or historical period perfectly embodying something that feels very important to us, our organization, our group, our society, or our era. Something that makes us say, "That's It! That captures it perfectly!" Energy comes from the collective unconscious when that sort of compelling flash of meaning occurs.

The collective unconscious has both an historical aspect and an archetypal aspect. The historical aspect consists of experiences accumulated by human beings throughout history: these experiences have entered, and are conserved in, the collective unconscious of human kind. Archetypes are dynamic principles that organize their manifold elements. They are irrepresentable in themselves, but have effects that make visualizations possible.

Ervin Laszlo, 2006

Research conducted by Jim Dator and his associates at the Center for Future Studies at the University of Hawaii has led them to articulate four scenario archetypes, four fundamental views of the future that appear to transcend history and culture [Dator 1998]. Although they recognize that any categorization necessarily simplifies the rich array of possibilities, they have found that these scenarios represent the predominant ways people think about the future. These four basic images illustrate sharp distinctions between alternative views of future possibilities.

We have seen and worked with a number of typologies of the future, but we find this articulation of primal images valuable because…
> these images recur across many cultures.
> they illustrate stark contrasts from one to the other.
> they suggest possibilities that people may otherwise overlook.

By using these primal images as a set related to one another in these two dimensions, we can see...
> how the images are formed and validated by those holding them.
> how several scenarios can occur simultaneously.
> how we might move from one reality to another over time.
> leverage points where human choices could help to trigger or to avoid a shift into another reality.

These archetypal scenarios allow us to delve into four distinct future "worlds" to experience their uniqueness and learn from them. The essential characteristics of each scenario are shown in the following graphic:

Exploring

	Abrupt	
COLLAPSE - Economic & technology decline - Socio-political disruption - Environmental degradation - Survival mentality - Anxiety & cynicism		**NEW REALITY** - Economic prosperity - Socio-political transformation - Tech & genomic breakthroughs - Radical new forms of belief, behavior & organization
Reactive		Creative
STATUS QUO - Economic disparity - Socio-political nostalgia - Technology maturity - Fear of change - Traditional forms of belief, behavior & organization		**DISCIPLINE** - Economic investment - Socio-political idealism - Will to sacrifice - Technology growth - New forms of belief, behavior & organization
	Gradual	

We now take a closer look at each of the scenario archetypes and the interactions among them.

Exploring

☐ Maintaining the Status Quo
[Gradual change / Reactive response]

EXAMPLE: In **Status Quo,** we would take an aspirin to relieve pain without addressing the cause of the pain. The pain may worsen when the effect of the aspirin wears off, and more potent medications may be needed to produce the same pain-reducing effect. Meanwhile, the underlying cause of the pain goes untreated. Ongoing attempts to mitigate the pain may result in small oscillations that keep it under control or perhaps increasingly larger oscillations of pain and relief.

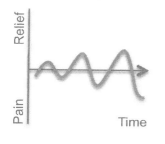

"We're on a long growth trajectory and that proves we know what we're doing. It's always worked in the past and will work in the future; there's plenty of blue sky ahead. Stick with the tried and true. We've faced challenges before, and we can handle anything that crops up. We're terrific problem solvers and have confidence that we can fix anything that comes our way."

In Status Quo, responses to critical uncertainties take the form of stop-gap measures intended to return to the way things have been in the recent past. It is a problem-solving mentality. The focus is on what is *not* wanted. There is strain and effort to keep the environment static and unchanging. The "fixes" in this scenario – while they may be necessary and expedient – are temporary because they don't address the fundamental issue. They mask the problem and may alleviate concern about it, but they do not resolve it. It is quite possible, though, that an organization or a social system will be able to "muddle through," maintaining the status quo for a *long* time.

Americans have long had an unswerving belief that technology will save us — it is the cavalry coming over the hill, just as we are about to lose the battle. And yet, as Americans watched scientists struggle to plug the undersea well [in the Gulf of Mexico], it became apparent that our great belief in technology was perhaps misplaced.

Elisabeth Rosenthal, 2010

This does not imply that only new solutions are fundamental or that anyone wanting to turn back the clock is seeking a "quick fix." Some new approaches have had unintended consequences. A fundamental solution may be to return to past practices – examples include, micro-manufacturing [batch and craft production of products] and using locally available materials for construction.

Because they were not addressed, the effects of Critical Uncertainties will reemerge, sometimes stronger [see the graph above]. Over time, this approach of relieving the symptoms diminishes the system's ability to address fundamental issues and may result in a sudden shift into the Collapse scenario.

Perhaps we never appreciate the here and now until it is challenged.

Anne Morrow Lindbergh
1906 – 2001

Exploring

☐ Exercising Discipline
[Gradual change / Creative response]

"The world is changing, and we need to change too. We can't depend on having all the resources we need when we need them – skills, raw materials, food, energy, clean air and water. We all have to pull together to try something new. Let's be willing to experiment! We can't apply yesterday's solutions to tomorrow's challenges. We can't wait; we need to take steps now."

In the Discipline scenario, people see critical uncertainties as opportunities to do things differently than we are currently doing them and hopefully better. It is called Discipline because collective discipline is required to change. Status Quo is characterized by the desire to fix problems. Discipline, by contrast, is marked by a desire to create. People go toward something in an open, hopeful, loving and communal spirit. We aspire to win – in a spirit of joy and fun mixed with urgency and necessity. The focus on creativity is a powerful force. Ideas increase exponentially.

Discipline is facing the source of pain and treating it effectively, even if the treatment requires time, effort and sacrifice.

In Discipline, we commit ourselves to a course of action because we believe it really matters. Our actions address the fundamental issues in lieu of symptoms. Fundamental action is likely to take longer and cost more than symptomatic solutions. The effect of our efforts may not be immediately obvious; it may seem as though nothing is happening despite all of our hard work. Things might even get worse before they get better. Delay in seeing the fruits of our efforts could lead to a redoubling of those efforts – perhaps resulting in an overshooting of the goal. Or it may result in discouragement and a loss of commitment.

A fundamental solution to a problem might very well be to return to past practices or to simply stop doing something that is causing a problem. For example, we could work to assure ourselves that the water from our taps is clean and pure in lieu of buying bottled water which increases our dependency on oil-based plastic products and adds to overflowing landfills.

Real, incremental changes could lead, over time, to a New Reality scenario.

Exploring

☐ **Stepping into a New Reality**
[Abrupt change / Creative response]

"Did people really think that humanity as we knew it in the twentieth century was the culmination of evolution? Probably. After all, they thought that Web 2.0 was the ultimate in technology. But our nanotech brain enhancers today are amazing! Brain wave communication puts me in touch with anyone any time. I do my best creative work while deeply asleep. It all seems so natural, like it's always been this way."

The Big Bang may have been the initial **New Reality** scenario. Suddenly, our universe appeared! The acceleration of technical innovation has placed us again at "the edge of change." Writer Vernon Vinge [1993] has advanced the idea that super smart computers might soon outstrip human intelligence. Is this concept, known as The Singularity, just the stuff of science fiction, or is it a possibility? Proponents of "conscious evolution," like Barbara Marx Hubbard [1998], anticipate the "emergence of a universal humanity capable of co-evolution with nature and cocreation with Spirit."

A New Reality can take many forms, both radical and subtle. Radical advances in nanotechnology, biotechnology, information technology and genomics may very well result in new life forms or new types of intelligence, leading to a dramatic transformation of the workforce, organizations and the nature of work. Androgyny is an example of a subtle form of a new reality. For eons, gender and sexual orientation have been central to the way in which humanity defines itself. If those distinctions cease to be important or relevant over time, the nature of social relations in the work environment would be significantly transformed, even if we can't now understand the precise implications.

In the animated film *Waking Life* [Linklater 2001], a young man floats in and out of philosophical discussions, giving a sense of what telescopic evolution might feel like – where something is envisioned and can be immediately experienced. Increasingly, what has seemed fantastic has to be considered more seriously. The gap between today and a far-off tomorrow may be shrinking.

It takes only a tiny group of engineers to create technology that can shape the entire future of human experience with incredible speed.

Jaron Lanier, 2009

63

Exploring

☐ Experiencing Collapse
[Abrupt change / Reactive response]

"In retrospect, we could have seen it coming – did we need to let it go this far? It feels like we are in free fall, but it has to stop somewhere. We can reorganize what we have left and start again. We've taken some hard knocks before. We know how to survive in the toughest of times. Some really good things might come out of this crisis!"

In a Collapse scenario, financial, social and/or environmental systems become dysfunctional. Carrying on life as we knew it is no longer possible.[7] Collapse is characterized by desperation, a hunkering down, bartering and new forms of currency, blame for those held responsible and a tribal-like isolation of groups. To avoid further loss or pain, self-protective instincts dominate: fear and competition lead to an unwillingness or inability to empathize with the plight of others.

A Collapse scenario might entail:
- Demand for fuel, food, water and other essentials outstripping supply
- Economic meltdown and new forms of currency
- Toxic environments causing physical and psychic pain; diminished bio-diversity.
- Global climate change affecting where and how people live and work
- Fierce, relentless competition and hostile takeovers
- Disengaged employees, partners, investors and customers
- An explosion of long-festering, neglected problems
- Civil disorder; increased crime and terrorism
- Ineffective and/or corrupt governments

Do we need to be in crisis before we are motivated to change? Collapse is probably the worst time to effectively implement fundamental solutions. Thinking becomes entirely short term and immediate. We are in panic mode, simply trying to survive.

> In **Collapse**, Jared Diamond [2005] defines an extreme collapse as "a drastic decrease in human population size and/or political/economic/social complexity, over a considerable area, for an extended time." He recognizes milder forms of collapse, such as the rise and fall of fortunes, restructuring, takeover of one group by another without change in the size or complexity of the whole, decline linked to another's rise or the replacement / overthrow of one governing body by another. Diamond identifies the forces leading to Collapse as:
> - *Damage people inflict on their environment*
> - *Climate change because of changes in natural forces*
> - *Decreased support by friendly neighbors*
> - *Response to problems*

---

[7] A "collapsitarian" site, www.postpeakliving.com, seeks to prepare us for life beyond collapse with courses such as *Navigating the Coming Chaos of Unprecedented Transitions* and *Sustainable Post-Peak Livelihoods*.

Exploring

- **The Scenario Game Board**
  The two-dimensions form a grid of the Scenario Game Board that reveals the structure connecting the scenarios. At a very high level, the structural dynamics connecting the scenarios looks like this...

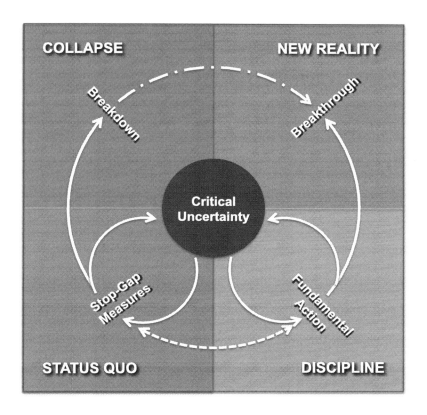

The Scenario Game Board
©2010

Here is how it works. As we've seen, a critical uncertainty is an event that would have a significant impact on the Decision Issue under consideration. We can use the Game Board to learn about the dynamic field surrounding any critical uncertainty. To get started, we place one critical uncertainty in the center of the Game Board.
  ‣ Abrupt change occurs above the center line; gradual change below.
  ‣ Reactive responses occur to the left; creative responses to the right.

Each quadrant, representing a scenario, is connected to the others.
  ‣ In Status Quo, stop-gap measures are applied.
  ‣ In Discipline, fundamental solutions are undertaken.

And, like one more grain of sand...
  ‣ In Collapse, one too many stop-gap measures results in breakdown.
  ‣ In New Reality, a series of fundamental solutions lead to breakthrough.

Exploring

The interconnections between the scenarios indicate their relationship and how one might lead to another.

The dashed line between Status Quo and Discipline indicates that these scenarios are intricately linked and rarely occur independently. In many cases Status Quo and Discipline must coexist. One or the other may predominate, and the balance between the two may be constantly shifting. It takes a great deal of Discipline to move from Status Quo. Failing to muster such Discipline could result in over reliance on stop-gap measures to maintain the Status Quo. Leaders have the difficult challenge of simultaneously managing within both Status Quo and Discipline. Status Quo products and customers provide the resources to support the Discipline work required to innovate new products and solutions to move their organizations into the future. One Anticipatory Leader calls this difficult challenge of operating in both Status Quo and Discipline being "ambidextrous." He uses the term to describe the ability of an organization to live with change

The dotted/dashed line from Collapse to New Reality indicates that, once in a Collapse scenario, it is extremely difficult [perhaps impossible] to leap to a New Reality. Many people believe that "It takes a collapse to bring about real change," but we hold that this outcome is the exception rather than the rule. For example, Rome was once the dominant power in Europe and the Middle East, holding sway over vast territories for centuries. When it collapsed, power fragmented and shifted to other players on the world stage. Many economies stagnated. It took Britain approximately 1,000 years following the collapse of the Roman Empire to get back to the level of development it had enjoyed in the Roman era [Ward-Perkins 2009].

The Game Board is constructed around each of the critical uncertainties. This systems perspective allows us to recognize the powerful dynamics in play that are driving behavior.

- **Being a Player**
  Neither the present nor the future fits into neatly defined boxes. These scenarios could occur simultaneously in different places. Some parts of the world may be experiencing a New Reality with a rapid series of technological breakthroughs while other places may be experiencing a Collapse of their natural, economic and social environments.

  The scenarios could occur sequentially in the same place, moving from trying to hold on to what exists in Status Quo to developing more sustainable practices and technologies in Discipline. In fact, Status Quo and Discipline often coexist – retaining the best of what is while simultaneously seeking fundamental solutions [new or old] to entrenched issues.

  The Game Board allows players to move from any quadrant to any other – or to be in several at once in different locations, in different domains and under different circumstances. It is a presentation of a dynamic system, a way to try out possibilities.

  You may notice a correlation between these scenarios and the approaches to change discussed in the Introduction:

  ▸ **Status Quo → Fix it**
    We like where we are and want things to continue. We may be making ourselves vulnerable if we naively project past trends into the future. Rosy circumstances

Exploring

make us think it will always be this way. Similarly, bad breaks can trigger a gloomy outlook.

▸ **Discipline → Try something new**
We can see that things are about to change, and not for the better. We need to do something fundamentally different. It will take a lot of sacrifice, creativity and persistence. Even though we may not see results right away doesn't necessarily mean that our efforts are ineffective.

▸ **Collapse → Let it collapse**
The foundations of our world are falling away. It may take a long time to recover lost ground, if recovery is even possible.

▸ **New Reality → Leap into a new reality**
We leap the abyss onto a new growth curve and hope that it will be substantial enough to sustain us.

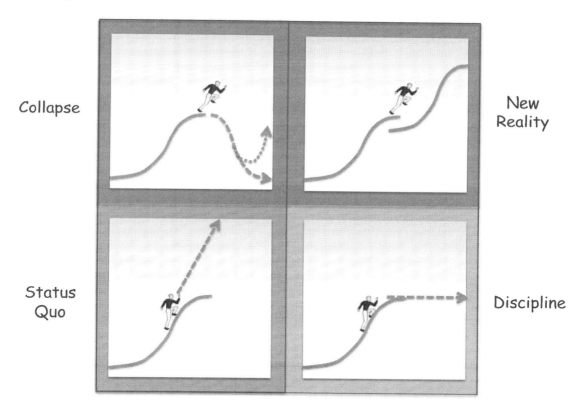

The structure of the Game Board shows us why, over time, individuals, groups, organizations and societies move from one scenario to another. Perhaps Collapse would propel us into a Discipline scenario; or it may make such a transition much more

Exploring

difficult or impossible. Like the myth of the Phoenix[8] rising from the ashes, Collapse may contain germs of renewal that could lead to Discipline or New Reality.[9] Some people believe that Collapse is necessary to get our collective attention and move us to critically needed action. In our view, organizations are in a much better position to change if they anticipate the full range of possibilities and make strategic moves to transition from one scenario to another while there is time and energy to do so in a carefully thought-out manner. In this way, we may be able to forgo the actual experience and severe consequences of a Collapse scenario.

The Game Board empowers players by providing a framework to think deeply about their situation, their assumptions and their options. The set of scenarios represents a wide range of possibilities. None of these scenarios is inevitable. And yet, none of them can be safely dismissed; each one needs to be thoughtfully considered.

The future won't match any of the scenarios exactly. Wild cards beyond anyone's ability to anticipate certainly do spring surprises [Petersen 1997]. However, as a set of distinct plausible futures, the scenario archetypes provide a context for thinking through large, complex, strategic issues. When an organization can visualize the causal structure surrounding its Decision Issue, it has a powerful analytic tool. Grasping the validity of these widely divergent futures by developing a sense of their structural interconnections is an important achievement in its own right.

In this design manual, we apply these four archetypal futures to the creation of life-sustaining organizations. Recognizing an organization as a living system in continuously changing relationship to and interaction with its employees, its suppliers and customers, its social and political context and to nature, leaders incorporate all these possibilities—both pleasing and shocking—into their strategic analyses.

Anticipating all of these scenarios as real possibilities, organizations are in a better position to make choices that will move them from where they are to where they want to be. They become high-leverage players in a high-stakes game.

We now go one level deeper to appreciate powerful, hidden drivers of structural behavior... mental models. Leaders who grasp the necessity of being life-sustaining recognize and challenge the assumptions and constraints of their own mental models in order to be open to seeing the broadest range of possibilities.

---

[8] A Phoenix is a mythical bird with a 500- to 1,000-year lifespan. Near the end of its life, it builds itself a nest and then ignites it. Both nest and bird are reduced to ashes, from which a new, young phoenix arises, born anew to live again.

[9] This myth is an allegory of renewal that reminds us of the story of monks, isolated in Ireland, who preserved much of Western civilization by carefully copying and illustrating manuscripts throughout the Dark Ages following the end of the Roman Empire. See Thomas Cahill, *How the Irish Saved Civilization.*

◎ Exploring Deeper

Below structure, at the very bottom of the dive, concealed among the silt, the weeds and the bedrock, lie our most deeply held beliefs, our values and the myths that frame our thinking. Beliefs are powerful elements of structure. They are so deeply ingrained in our cognitive and emotional processes that we are often completely unaware of their existence, let alone their implications. They help to explain why people make the decisions that they do. Mental models, myths and metaphors have a tremendous effect on human perception and, therefore, human choices. As such, they strongly influence the behavior of any system under consideration.

Mental models are both our short-term perceptions of the world and those that we hold in our long-term memory. Myths and metaphors are cultural representations of long-term mental models. Myths are deeply held stories that trigger emotive responses [like the myth of a charismatic savior]. Metaphors are inventive, allegorical language used to communicate some cultural truth or insight.

In our everyday awareness, we constantly rely on perceptions and assumptions to make sense of the world around us. We absolutely need them to be able to function. However, these short-term mental models accumulate over time and may gradually become firmly fixed in our thinking as deep-seated beliefs. After a while, we forget the original circumstances that caused us to make a particular judgment and simply rely on what we "know." The collective beliefs of a culture or a community become myths that reinforce and shape our individual perceptions.

For a *very* interesting test of perception see..

viscog.beckman.illinois.edu/flashmovie/15.php

These thought forms occupy a deep emotional place in personal and collective consciousness, touching our hearts and feelings. The substructural level shapes both what we see and how we see. If our mental models don't recognize that something is possible, we literally cannot see it – even when it is in front of our eyes [the video cited in the box is a powerful example of this phenomenon]. The fundamental stories that we subconsciously hold drive our views of the past, the present and the future. Beliefs about whom we can trust, and under what circumstances, are deeply engrained in the psyche of individuals and groups; they remain largely unexamined in the work environment. Assumptions that align with the Status Quo scenario are quick and easy choices because they "stay within the lines" of how we see reality. In this situation, we don't feel a need to challenge our perceptions.

When we say someone is a warm person, we do not mean that they are running a fever. When we describe an issue as weighty, we have not actually used a scale to determine this. And when we say a piece of news is hard to swallow, no one assumes we have tried unsuccessfully to eat it.

Drake Bennett, 2009

Exploring

There are many mythic, emotionally laden forces that shape how we view our worlds. We may be using filters that influence our views of....

freedom	↔	control
individual	↔	collective
self reliance	↔	family, community support
evolution	↔	creation
human dominion	↔	humans as part of nature
distinctions	↔	egalitarianism, oneness
free will	↔	destiny, God's will
reason, logic	↔	emotion
art	↔	science
linearity	↔	randomness
shareholder value	↔	social impact
one life	↔	reincarnation
materialism	↔	spirituality
simplicity	↔	complexity

These descriptors aren't mutually exclusive, but we tend to think of them as if they were. For example, we used to think that quality and low price were mutually exclusive. We had to pay dearly if we wanted design, craftsmanship, performance and reliability. Then the Japanese auto industry showed that this assumption is not always true [and then, with its recent problems, Toyota showed us that our new myth wasn't totally accurate either!].

Awareness of these mythic forces allows us to include their impact in our analysis; they can dramatically shift outcomes in ways that often seem mysterious. If we ignore them, we miss an opportunity to step back from those connections we take for granted and look at critical issues with a fresh mind. This practice leads to clearer perceptions and insights, making us better futurists.

Cultural legacies...persist, generation after generation, virtually intact, even as the economic and social and demographic conditions that spawned them have vanished, and they play such a role in directing attitudes and behavior that we cannot make sense of our world without them.

Malcolm Gladwell, 2008

Exploring

◉ Executive Briefing: Facts

Moving from events to patterns to structure and on to mental models, myths and metaphors, the Exploring phase covers a lot of territory. Rigorously identifying relevant facts, their patterns of interactions, the systemic structure of the pattern sets and the underlying beliefs and assumptions is an exhilarating process, leading to many insights and new, creative ideas. Organizations engaging in this process will come to understand how big-picture forces affect their ability to become life-sustaining.

The Team has been communicating with its constituents from the outset of the initiative. Everyone in the organization has had the opportunity to participate in the conversation, providing ideas and comments. At the completion of the Exploring phase, the Team presents the findings to the organization's formal decision makers. Because the executives are either participating on the Team directly or have one or more of their people directly engaged, they have a sense of ownership for the process and its outcomes.

The Team will most likely be highly enthusiastic about its work. Team members need to carefully describe their process to the executives, taking care to explain any terms that may sound like jargon to those who have not been involved. The goal of this meeting is to establish a baseline of common knowledge and information.

The executives will recognize many of the forces in play in the organizational environment but may not have thought through how they interact and the possible consequences. The Team may be deflated if the executives respond with comments such as, "That's obvious," "That's nothing new" or "We've known that all along!" Members should remember that the obvious is frequently forgotten and what was once known has to be seen again.

> We shall not cease from exploration
> And the end of all our exploring
> Will be to arrive where we started
> And know the place for the first time.
>
> *T.S. Eliot, 1942*

Just as importantly, the Team will explore territory that is, at least in part, somewhat unfamiliar. Even though many executives have a finely honed strategic sensibility, the environment may have changed in subtle and unanticipated ways since the last time they looked. It is very possible that the Team will have discovered critical uncertainties that the executives had not noticed or had not fully realized their importance. Either way, the group as a whole needs to explicitly acknowledge and accept the facts developed during the Exploring phase as a firm foundation for a dialogue about the organization's future.

Following the Executive Briefing, the Team communicates the facts established in Exploring through the Playbook, documenting the progress and agreements so far. The Team is now ready to move into the next phase of the work – Discovering. They transition from strategic thinking to strategic analysis.

Discovering Options

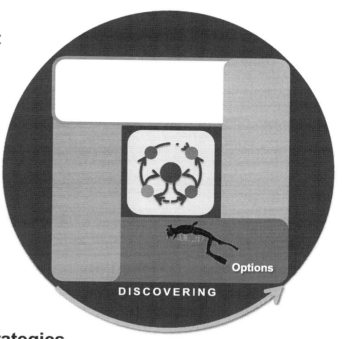

- **Role of Leaders: Strategist**
  - Your Experience
- **Analyzing Future Possibilities**
  - Strategic Insight
  - Success Leads to Vulnerability
  - Thinking the Unthinkable
- **Articulating Scenarios**
  - Creating the Structural Dynamics Model
- **Living in the Future**
  - The Organization in the Scenario
  - Naming the Scenarios
  - Scenario Narratives
- **Developing and Testing Strategies**
  - Effective Action in Each Scenario
  - Stress-Testing Strategies
  - Creating a Scenario Matrix
  - Robust Strategies
  - Contingent Strategies
  - Testing Existing Strategies
- **Executive Briefing: Options**

Discovering

Role of Leaders: Strategist

In Exploring, we identified the forces in play in the organizational environment and used that information to develop a set of structural diagrams for each of our critical uncertainties. This process has given us a view into the dynamics surrounding the Decision Issue under consideration. In Discovering, we develop the structural analysis into a set of alternative future scenarios. These scenarios provide the basis for developing and testing strategies in a wide variety of conditions without the risk associated with taking such actions in the real world. Without the pressure and constraints of a live performance, the organization can test, observe, rethink and test again. Learning the skills of strategic thinking is an invaluable part of the Structural Dynamics process.

> Every profound innovation is based on an inward-bound journey, on going to a deeper place where knowing comes to the surface.
>
> W. Brian Arthur
> Santa Fe Institute

As the life-sustaining initiative progresses, Team members develop a sense of ownership for the results. The focus now shifts to strategic analysis. By comprehending the dynamics in play, strategists are able to identify interventions that will turn the energy in the system toward the outcomes the organization is seeking to achieve. They work in concert with this energy like martial artists, taking the forces that exist and redirecting them toward informed action. They formulate their systemic insights into strategic options that yield life-sustaining organizational results. The Team then communicates their understanding throughout the organization.

Marlon Brando
1924 – 2004

Structural Dynamics provides a context in which these shifts in thinking and awareness are more apt to occur. Like a director using method acting,[10] leaders as strategists set the context for insight by getting so deeply "in role" in the future that they actually feel they are there. The Team leader encourages participants to suspend any disbelief they might have about their scenario and just consider what the world would be like "if" it occurred – after all, scenarios are "only stories." Developing and "living in" future scenarios represents an opportunity to experience the real possibilities inherent in these futures. The scenarios are distinct from current conditions, distinct from assumed future conditions and distinct from each other.

Because the scenario set is based on the causal analysis that the participants explored together, each scenario is equally plausible. Experiencing both positive and negative aspects of multiple futures with both heart and mind opens participants' receptivity to what might be. In this reflective environment, insights come naturally; Team members quickly recognize them and immediately integrate them into their construction of future possibilities.

---

[10] Method acting is a technique in which actors, such as Marlon Brando, aim to engender in themselves the thoughts and emotions of their characters to create lifelike performances. It can be contrasted with more classical forms of acting, in which actors simulate thoughts and emotions through external means, such as vocal intonation or facial expression.

Discovering

One of the skills developed in this process is experimental thinking about effective action. Practice leads to mastery. Musicians rehearse. Armies simulate battle conditions in war games. Athletes work out constantly. Great performances of all sorts require seemingly endless practice. In Outliers, Malcolm Gladwell reports that to be truly great in a field requires 10,000 hours or more of practice. This principle applies across disciplines to such diverse skills as debating, performing arts, design, engineering, improvisational comedy and firefighting – and it holds true both with individuals and with groups. While the precise action that will be required at any moment is unknown, the hours and years of preparation provide a sense of knowing how to perform as the situation requires. The more practice, the better the performance! Strategists need the opportunity to do the same, given that so much is at stake in their performances.

If I'm trying to sing something and I can't get it, I'm going to keep at it until I get where I want it.

Ray Charles
1930 – 2004

Strategic thinkers in organizations often don't get the opportunity to drill, rehearse and practice together. Perhaps that is why so many strategic plans fail to achieve their goals. The process of generating strategies by living in scenarios provides organizations the opportunity to take chances and to fail, fail fast and fail often without suffering the consequences. The organization investigates, tests and cross-tests a full range of options to identify a strategic direction that inspires deep confidence.

The ability to develop strategy by using both rational thought and gut feeling is a distinctive competence. In *Napoleon's Glance: The Secret of Strategy*, William Duggan [2002] posits that great strategists have an encyclopedic knowledge of particular fields. So, when something happens that resonates with what they have already considered, they are able to "glance" at it and draw connections that no one else sees.

Organizations as well as individuals can develop a strategic sensibility. *The Wisdom of Crowds* [Surowiecki 2004] makes the case that groups may be better at strategy development than individuals. Other research indicates that groups might have more than mere wisdom; they may possess a genius that individuals can't approach, as in the instance in which a group of people working together solved a long-standing mathematics problem [Ellenburg 2009]. Thus, the skills of strategist and integrator combine to impart this wisdom throughout the Team to benefit the organization well beyond the duration of the Structural Dynamics process. Skillfully bringing people together yields better organizational thinking.

Organizations that plot strategy solely from numeric data or by responding to the issues of the moment are missing something fundamental – the serendipitous understanding that emerges when people find themselves in a shared context. In this context, strategic thinking goes beyond imagining future conditions to having an "experience" of them. This process

Discovering

can be a rich adventure with the prospect of unexpected discovery. Those who think together about the future engage in a life-sustaining activity. Teams of strategic thinkers who are able to immerse themselves in the future have a powerful knowledge-sharing experience that leads to shared understanding and enables insight. The opportunity for this kind of "genius" to emerge from the group makes the work environment highly attractive to creative talent.

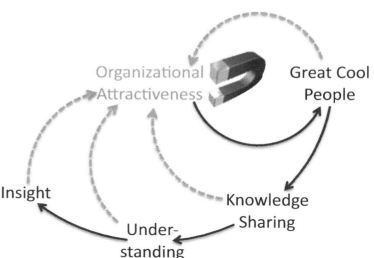

- **Your Experience**

    Use this self-assessment tool to consider your own skills, attitudes and values as a strategist. While these reflection questions are not an exhaustive inventory of strategists' skills and mental models, they may highlight strengths already present in your way of thinking and your leadership style.

Strategist Self-Assessment Tool

Are you...

☐ able to effectively use insights to produce desired results throughout an organization?

☐ at ease setting an overall direction for a group of implementers?

☐ taking a disciplined approach to focusing the attention of others beyond immediate, short-term pressures?

☐ constantly interpreting [and reinterpreting] how organizational reality is unfolding and what that means for the future?

☐ able to discern the system of forces at work in your field well enough that you foresee how events and outcomes may turn out?

☐ creating structures, e.g., processes, long-term investment-analysis activities, workplace design, location, policies, etc., that support strategic thinking in others?

☐ aligning the assets and processes of your organization to realize its strategic intent?

The questions may also cause you to pay more attention to other, perhaps underutilized, skills. Rank your skills in the strategic competencies presented here.

It can be helpful to remember and write down several specific instances in which you manifested the skill described or you missed an opportunity to do so. If you have demonstrated strengths in some of these domains but less so in others, you may want to look for opportunities to practice particular competencies.

☉ Analyzing Future Possibilities

The idea of using multiple views of the future for strategic analysis was developed by the military for war gaming in the 1940s. Two decades later, private and public organizations began adapting scenario thinking to organizational strategy development. Since then, scenarios have permeated the work of many futurists and strategists. We are now seeing a resurgence of interest in this powerful approach.

★ "In a highly uncertain environment, the advantages of scenario planning are clear: since no one base case can be regarded as probable, it's necessary to develop plans on the assumption that several different futures are possible and to focus attention on the underlying drivers of uncertainty."
>> "Strategic Planning: Three Tips for 2009," *The McKinsey Quarterly*, 4/2009"

★ "For 2009, we have sharpened our strategic processes and scenario planning ... Each of our businesses has set up a process to identify the 'naysayers' in each of our industries to make sure their voices are heard inside GE. From top to bottom and across GE, we must and will listen more critically and respectfully to each other."
>> "President's Letter to Shareholders," *General Electric's Annual Report*, 2008

★ "It's clear that major financial institutions had not been exhaustive enough in understanding how their portfolios and businesses would fare in a variety of scenarios. Scenario planning – relentlessly asking 'what would happen if?' – is a particularly essential discipline in the financial industry. Too often, leaders assume that the future looks like the past and build their business around assumptions that don't hold up. Leaders in the financial industry clearly didn't model just how vulnerable their businesses were to very plausible scenarios, and thus failed to curtail or counter certain risks."
>> "Building Better Wall Street Leaders," *Washington Post*, 9/19/2009

Structural Dynamics integrates systems thinking with scenario analysis. This integration ensures that the resulting scenarios emerge from a careful analysis of causal relationships present in the system. As a result, they are much more than fantasy or "imagineering" – the scenarios have internal validity. They make sense because their logic is based on the interdependent relationship of the forces and variables. Because they are based on real dynamics, they are capable of leading to strategic insight.

[Scenario analysis grounded in a causal analysis of dynamic complexity] creates the best quantitative representation of continuous variables that describe the future state.

>> "The current state of scenario development: an overview of techniques," *foresight*, Bishop, Hines and Collins, 2007

Discovering

- **Strategic Insight**

  Insights are the core of strategic analysis. They are more apt to occur when the issue under consideration is on the periphery of our thinking, not in our direct focus. Insights occur when we are in a reflective state and able to listen, absorb and consider; when we are present in a state of being rather than doing; when we make the time and space to think expansively.

  Like the calming effect of vigorous exercise, insights often emerge after the conscious mind has worked hard to sort out the facts and consider alternatives. Something new may come forward when the conscious mind relaxes and the subconscious continues to mull over the issue. Having an insight is effortless; like discovering a truth that was just waiting to be found, to be remembered. This state of being in the moment while reflecting on an important question without the noise of internal criticism is similar to Christian grace, Taoist *qing,* Buddhist cessation, Islamic opening-the-heart, Jewish *kavanah* and Hindu wholeness or oneness.

  Things seem to fall into place in a way that makes more sense than they had before. It is as if we suddenly grasp the essential nature of something. Insights have a powerful quality of not seeming to belong to one person – we have a sense that we've hit upon a truth that has universal resonance. For this reason, blinding insights might later seem deceptively simple or obvious. Insights shift thinking in a second. There is no need for "retraining"; old beliefs are gone. It is nearly impossible to remember how we thought about something before the insight occurred [Keifer, Charbit, and Manning 2009].

- **Success Leads to Vulnerability**

  Scenarios are an excellent way for highly successful organizations to defend against complacency and arrogance. When they believe they have "cracked the code" and are doing everything right, they become vulnerable. They fall prey to the trap of relying on past and present trends without testing their assumptions about the future. Years of experience take precedence over a fresh look at a situation. Believing in their own infallibility, organizations become blind to shifts in the organizational environment and open themselves to significant risk. Taking strategic action based on a single view or theory of the future can be fatal.

  Lulled by Success

  A company in the business of distributing cable and wire for high-voltage transmission has enjoyed steady growth and prosperity throughout its 75-year history. A proposed move from their inner-city location to the suburbs caused leaders to consider what they might need to do differently in the new facility. Although they were aware of many workplace concerns that the move would generate, they assumed that the demand for their products and their expertise would continue unchanged. They regularly read trade journals and tended to limit their view of the future to fluctuations in the price and availability of raw materials and inventory. Developments such as wireless transmission[11] or local generation of electricity remained

---

[11] A concept of Nikola Tesla [Cheney 1981]

firmly outside their image of the future. Leaders didn't take into consideration that advances in technology, global competition and changing customer expectations could become important to this company within a relatively short period and certainly within a 10- to 25-year time horizon. Their way of thinking illustrates the Status Quo archetype. The company's leaders felt no need to consider such external possibilities because they had never done so before and things had worked out just fine.

- **Thinking the Unthinkable**

Prepared to Leap

Phenomenal change is afoot in nearly every domain, with powerful forces reshaping the future. Because they are out of an organization's control, people often omit them from thinking and planning; this oversight ignores the "elephant on the table," the big issues that people collude to keep from discussing. Many organizations fail to anticipate where technology and science are taking us. If New Reality or Collapse occurs, these organizations will be completely unprepared. Never having thought about the organizational consequences of abrupt change, they are like deer caught in the headlights, completely paralyzed in shock and fear.

In the 1980s, Digital Equipment Corporation was the second-largest computer company in the world after IBM. DEC's founder and president, Ken Olsen, had made a series of prescient decisions regarding future trends in hardware, operating systems, applications, microchips and consulting services – he'd been right on virtually every bet-the-company decision he had made. The company had sustained double-digit growth for nearly 20 years. Thinking in terms of DEC's refrigerator-sized minicomputers and their peripheral terminals, printers and storage devices, Olsen infamously said, "Who would want a computer in their home?" At a critical shift in the history of computer technology, Digital wasn't thinking the unthinkable. The organization that pioneered the industry shift from mainframes to minicomputers failed to successfully leap into the personal computer era.

Before Europeans discovered Australia, we had no reason to believe that swans could be any other color but white. But they discovered Australia, saw black swans, and revised their beliefs. My idea...is to make people think of the unknown and of the potency of the unknown, particularly a certain class of events that you can't imagine but can cost you a lot: rare but high-impact events.

Nassim Nicholas Taleb, 2007

Discovering

Articulating Scenarios

We are now ready to create compelling images of future possibility. To do that involves several steps:
- creating a Structural Dynamics Model by linking common variables in the diagrams.
- living and working in the future in four distinct scenarios depicted in the model.
- naming the scenarios and crafting narratives.

- **Creating the Structural Dynamics Model**

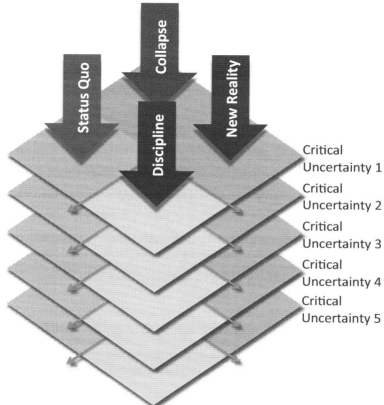

The Structural Dynamics Model

A scenario, like life, has a number of layers that affect one another. So far in the Structural Dynamics process, we have diagramed the dynamics of several critical uncertainties one at a time. We often find that several variables are present in more than one diagram, e.g. these variables have a causal relationship to two or more critical uncertainties. Thus, the critical uncertainties are connected through these common variables.

By considering how the system diagrams of the critical uncertainties interact, we develop dimension and depth in our thinking. Imagine the diagrams of the critical uncertainties as stacked layers. Now shift perspective and look down through the layers. By noting where variables repeat on two or more levels, we make connections between them and

Discovering

sense how the structural diagrams affect one another. We can trace how a change in a common variable affects more than one critical uncertainty. The variables, including the critical uncertainties, can magnify, refract, diminish and occlude one another in complex ways. The richness inherent in each scenario begins to emerge.

> **EXAMPLE: Strategic Investments**
>
> To illustrate the process of developing a Structural Dynamics model, let's consider this Decision Issue:
>
> *"What investments should this organization make to shape public policy?"*
>
> Let's say the time horizon is 15 years. Here are critical uncertainties related to the issue:
>
> - The impact of rising levels of atmospheric CO_2
> - The degree of public influence on private enterprise
> - The nature of international agreements governing nations and institutions
> - Women's social, political and economic position relative to men's
> - The degree to which the world's population is literate, skilled, educated and engaged in lifelong learning
> - The extent to which cities are cohesive, thriving centers of culture and attainment
> - The availability of fresh water, clean air and natural resources
>
> During the Exploring phase, the Team would create a structural diagram for each of these critical uncertainties. Participants then imagine these diagrams stacked one upon the other. The Team looks down through the layers of the stack, noting the dynamics in each quadrant and how the layers interconnect. Members of the scenario group place themselves 15 years into the future. Using the information from the interconnections of the structural diagrams, the group develops an image of their scenario world and looks back to the present to see how things got to where they are.

The participants join one of four groups, each focused on a single scenario.

→ These scenario groups are carefully formed from the members of the Team in a way that maintains as much of the Team's multifaceted diversity as possible.

→ Participants with strong belief in one particular scenario are assigned to a different scenario so that they can experience that possibility.

→ If a participant clearly does not believe that a particular scenarios is possible, this is an opportunity for that person to 'live" in and experience that world.

→ The groups work in separate areas within one large room. This creates a buzz of excitement and keeps the groups somewhat aligned with each another in terms of timing.

→ The space should be large enough so that each group has acoustic privacy and participants don't need to speak loudly to be heard.

Discovering

→ Each group looks at the dynamics in its own scenario quadrant. For example, the Discipline scenario group considers the interactions of all the critical uncertainties within the lower right quadrant of the Scenario Game Board.

🌐 Living in the Future

As a first step, the group creates an sense of what life is like for people in its scenario world. The members put themselves in that setting and experience the emotions and sensations of that world. At this point, they are not thinking about their organization. They consider questions such as:

"What is it like to live in this world?"
Each group details the ways in which its world affects people's lives: the home environment; the self; relationships; community; the natural environment; career contributions; physical, emotional, mental and spiritual health; hobbies and interests; the nature of work; child-rearing practices; the quality and type of public institutions and community resources; modes of transportation, etc. Participants consider how the world affects their personal choices and the advice they give their kids. They describe personal life in this world in full color.

"What is it like to work in this world?"
The groups consider where they work, how they communicate, how their performance is measured and how they are compensated. They discuss how much they find their work meaningful and fulfilling.

"How did this world come into being?"
To determine how their world evolved from where it was to where it is now, each group develops a chronology of events. For example, the Collapse group might find that a combination of pandemics and conventional warfare got them to where they are.

- **The Organization in the Scenario**
  Once the groups have a strong sense of life in the scenario, they develop a vivid image of their organization in the scenario world and "experience" working in that environment. The following questions can be used to guide the discussion:
  ▸ What products and services does the organization provide?
  ▸ Who are its customers, suppliers and competitors?
  ▸ How international or local is the organization's activities?
  ▸ What does the workforce look like? How diverse is it?
  ▸ How does work get done?
  ▸ What is exciting and engaging about work?
  ▸ What kinds of tensions show up between people?
  ▸ Where do people work: in offices, at home, at various remote locations, etc.?

  In each of these scenarios, consider:
  ▸ What are the strengths and vulnerabilities of the organization in the context of the scenario?

- What attracts or repels creative talent?
- Are there any distinctive elements of the organization's work culture?

Creating a sense of life, work and the organization in each scenario is an illuminating experience that generates new information, understanding and insight. To document and communicate the essence of the scenario, we encourage the creation of images, skits, posters, videos [e.g., a characterization of family life, a news report, a scene from a popular drama, etc.]. Visuals help others to grasp the nature of life and work in the scenario. These will be used in the next session while developing strategy, at the Executive Briefing and to communicate within the organization. Through these visuals, the places in the room where the scenario groups have been meeting now take on the character and appearance of their worlds.

- **Naming the Scenarios**

  With an image of the world and the nature of the organization in this scenario, each group then gives its world a short, catchy name that captures something fundamental about it. For example:
  - ★ Today Tomorrow [Status Quo]
  - ★ One for All and All for One [Discipline]
  - ★ Co-evolution [New Reality]
  - ★ Gaia Shrugged [Collapse]

  These names quickly become a shorthand within the organization, allowing complex ideas to be communicated quickly and simply.

- **Scenario Narratives**

  At the end of the session, the scenario groups review what they have learned about their worlds and capture the essence in bullet form. It is helpful to use the STEEEPA categories to ensure that the groups have considered all of these domains. During the break between sessions, each group will write a one- to two-page narrative that includes a description of life and work in the scenario and the events that caused this scenario to come about. If the summaries are longer, they become too cumbersome for their intended purpose of providing a context for decision making and action. Even in the following brief illustrative narratives, a richness quickly develops by looking at the interconnections of critical uncertainties related to the Decision Issue.

Discovering

EXAMPLE: Strategic Investments – <u>Continuing our Example from page 81</u>
Here are some brief outlines of the scenarios for 2025 that emerge using the interconnected critical uncertainties.

Status Quo
A perpetual state of anxiety affects people individually, and socially. Business, governments and academia maintain a command and control style that they know and trust, working long hours and relying on rules and regulations. When something unexpected happens, they believe someone is to blame, more control is needed and everyone needs to try harder. There is a pervasive confidence that technical solutions can and will be found in the nick of time to avert disaster [after all, the "hole in the ozone layer" problem was fixed]. When substitutes can't be found for scarce resources, stock-piling occurs which spawns conflicts. Social programs demanded in Europe and Canada are met with resistance in the United States which maintains its self-image of a country of individualists: most 20th century safety nets [social secuity, medicare, etc.] have been altered, abandoned or privatized. Women and minority groups have developed areas of expertise where they take leadership roles. People continue their educations throughout their lives to keep pace with increasing complexity in nearly every field. There are dramatic economic discrepancies between countries and within countries. It has become increasingly difficult to maintain a good quality of life in urban areas due to rising costs, over-crowding and crime. The affluent live in gated communities or other means of seclusion: their access to culture, events and other people is mostly virtual.

Discipline
Increased problems caused by rising levels of greenhouse gases in the atmosphere leads to international agreements mandating reduced emissions. Global corporations start cooperating with each other to develop new methods, new energy sources, and new technologies to reduce their collective impact on the environment. Micro-manufacturing allows companies to produce many of their products close to consumers, greatly reducing their carbon footprint. Small
farms spring up in unexpected places – such as urban centers. Many new materials are available, e.g., biodegradable plastics. The high demand for scarce talent has shifted the focus of employers away from gender, race, ethnicity, and other personal factors to simply the person's capability, opening up a range of opportunities for the formerly disenfranchised. People of all ages and types are attracted to urban centers for job opportunities combined with educational and cultural resources. Urban residents work collaboratively in their neighborhood toward continually improving the urban experience. This creates a rebirth of vitality and cities develop their own unique characteristics which further encourages an influx of residents and visitors. Rural areas, including some former suburbs, provide an easily accessible experience of nature.

Discovering

New Reality
The magnitude of catastrophic weather events finally convinced the G20 governments to act in concert: in 2015 it became clear that a significant change in ocean currents could happen with disastrous effects. Global concern for the planet led to unprecedented levels of cooperation. The international community began investing heavily in alternative energy sources. Eight years later, the world was largely free of its dependency on burning carbon for fuel. Droughts and hunger were eliminated by the ability to make water out of thin air, even in the most arid of atmospheres. Global cooperation led to a diminishment in the importance of nation states and regional conflicts. Most people now consider themselves to be citizens of the planet Earth; choosing to live wherever they want around the world. There have been no wars since the last of the nuclear weapons were dismantled five years ago and their fuel recycled for beneficial purposes. The massive change efforts spurred educational attainment and entrepreneurship. The focus of life on Earth is now on quality and sustainability. Everyone is well fed and sheltered but not at the extravagant scale that some had enjoyed at the turn of the millennium. That level of consumption is now viewed as immoral and socially unacceptable. Global social pressure has largely replaced the need for legislation and enforcement. Much of the world participates in a gift economy. No one holds onto more resources than they need to live comfortably.

Collapse
The global atmosphere changed much more quickly than scientists predicted even a few years ago. Rising sea levels have destroyed much valuable coastal land including many major cities. Docking facilities have become inoperable, putting a major brake on international trade and disrupting food supplies around the world. Those who grow food locally are inundated by starving people unable to pay the high prices. Power rests largely in the hands of vigilantes; private armies attempt to provide security for those who can afford their services. Scarce materials and resources are the source of conflict and warfare in the constantly shifting landscape. International agreements are ignored as each nation and social system scrambles to do the best they can for their citizens. Women provide subsistence incomes for their families; supplying needed goods and services while men provide physical and military protection for them. Education is the responsibility of the individual – accessing whatever books, people and other sources of information they can find. Cities have closed their doors to those wishing to find refuge within them. They struggle to provide services for their residents. Within these urban enclaves, upscale and ethnic neighborhoods guard their territory against outsiders. Plagues, famines, wars and natural disasters have caused the global population to plateau at six and a half billion people.

Discovering

💡 Developing and Testing Strategies

Once scenario group members have a good sense of the dynamics of their world, they develop and test possible strategies pertaining to the Decision Issue. Here the group has the opportunity to experiment, observe and reflect without the consequences of a real-world trial.

- **Effective Action in Each Scenario**
  Within the context of each of the scenario worlds, the groups develop and test strategies to create a life-sustaining organization. They develop several strategies for their scenario, each of which is consistent with that scenario and effective within it. Each strategy has a clear rationale in that context and manifests a continuing commitment to grow a life-sustaining organization. To illustrate, here are some proposed strategies for each scenario:
    + In Status Quo: To increase productivity while sitting in traffic, an organization provides autopilot devices, voice activated smart phones and wireless ear buds to all employees.
    + In Discipline: An organization financially supports the development of public transportation options, locates its facilities at transportation hubs and provides no on-site parking.
    + In New Reality: Ubiquitous holographic conferencing, allowing people to interact remotely in ways that are more realistic and compelling than anything previously available, has all but eliminated the need to be physically co-located.
    + In Collapse: The organization becomes a community resource by distributing excess electricity it generates at night to households in return for their excess capacity during working hours.

  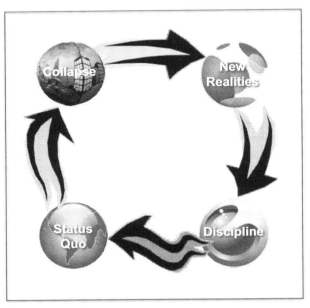

  Organizations need to identify and test strategies within their own unique set of circumstances and possible future conditions.

- **Stress-Testing Strategies**
  Strategies developed for a single scenario are "stress-tested" across the other scenarios to determine if the strategy works there, too. This process takes place experientially, as "ambassadors" from each scenario group sequentially visit the other scenario worlds in a round-robin or World-Cafe[12] fashion.

---

[12] See www.theworldcafe.com.

Discovering

Each visit to an alien world has these components:

1. Members of the scenario group tell the visiting ambassadors about life and work in their scenario world – they do NOT reveal their strategies. The scenario group uses the props and visuals they have made to portray the main characteristics of their world.

2. The ambassadors describe the strategies their organizations have developed for their own world, one at a time. [Note: Ambassador do not need to describe the world they came from.]

3. The group in the scenario world that the ambassadors are visiting responds with an assessment of how each of the ambassadors' strategies might play out in their world. [Note: When a strategy developed for a Collapse world, for example, is found to work in a Discipline-based scenario, it is often for very different and unexpected reasons.]

4. The ambassadors carefully record the results of the stress tests on the worksheet shown below. They note whether a strategy works in this alien world, it is a disaster in this world or it is neutral – it doesn't hurt but doesn't help. The ambassadors also note the reasons for these findings.

Ambassadors' Worksheet

Strategies \ Scenarios	Our Scenario World	First World Visited	Second World Visited	Third World Visited
	+			
	+			
	+			

5. When each strategy has been tested and the results recorded, the ambassadors travel to the next world and repeat the steps above.

Once the ambassadors have visited each of the other scenario worlds, they return to their own. There, the scenario group as a whole reviews the test results for their strategies in the other worlds. Do their strategies stand up in other scenario worlds? They enhance and clarify the notes in the Ambassadors Worksheets.

Discovering

- **Creating a Strategy Matrix**
  A strategy matrix is a powerful tool, in that the value of the strategies across the different scenarios visually "pops out." The groups record the scenarios in the columns and the strategies developed for each scenario in the rows. The strategies developed for a scenario are expected to work well in that world; this is shown with the "+" sign.

Strategy Matrix — **Scenarios**

Strategies

		Status Quo	Discipline	New Reality	Collapse
Status Quo	Strategy 1	+			
	Strategy 2	+			
	Strategy 3	+			
Discipline	Strategy 1		+		
	Strategy 2		+		
	Strategy 3		+		
New Reality	Strategy 1			+	
	Strategy 2			+	
	Strategy 3			+	
Collapse	Strategy 1				+
	Strategy 2				+
	Strategy 3				+
Existing	Strategy 1				
	Strategy 2				
	Strategy 3				

Using the data from the stress tests, the Team as a whole engages in a lively discussion about the merits of the various strategies. For each scenario, the group carefully reviews the supporting logic for the strategy, its implications and its consequences.

Based on this discussion, the Team fills in each cell in the matrix with "+" for works well, "-" for disaster/bad idea or "~" for neutral. They rigorously challenge each rating [+, -, ~] until there is a general agreement as to which symbol belongs in the cell. As this matrix is completed, it may reveal both robust and contingent strategies.

- **Robust Strategies**
  Robust strategies work across all scenarios. No matter what happens, an organization is making a good choice by investing in a robust strategy. They represent little risk because they do not depend on a particular set of circumstances to be effective.

  When the Team discovers robust strategies, they must carefully review and test all the assumptions they can identify upon which the effectiveness of the strategy depends.

Discovering

> **EXAMPLE:** A marketing firm develops a robust strategy to *"affirm and enhance the creativity of their workforce through a deeper appreciation of the arts."*
>
> ➡ In a Status Quo scenario, the organization benefits by supporting creative outlets for its employees that enhance the organization's visual and social environment, leading to a sense of shared achievement along with recognition of individual contributions.
>
> ➡ In a Discipline world, by placing an emphasis on creative expression, the employees have become extremely agile in prototyping new ideas, testing their viability and adopting or adapting those with potential.
>
> ➡ In a New Reality setting, art and design have become core operating principles. Innovation is the basic ingredient of the organization's value proposition. Imaginative virtual reality simulates and enhances real-world experience.
>
> ➡ In a Collapse world, the organization provides a haven and magnet for artistic expression of all types. It finds itself swamped by talented, qualified applicants. As a result, the firm is a standout in providing a unique experience to its customers in an otherwise dreary world.

- **Contingent Strategies**

    Contingent strategies work in some but not all scenarios. The value of having a rich repertoire of contingent strategies lies in the fact that the future will unfold along certain pathways and not others. Therefore, having a number of "If...then" strategies makes excellent sense, i.e., "If the future is shaping up in this way, then we'll put into action the strategies that work in those conditions." Contingent strategies should be captured in the Playbook to be used if and when events call for them.

- **Testing Existing Strategies**

    This part of the process presents a golden opportunity to test any strategies that the organization is using or plans to implement. Add them to the strategy matrix and test them across each scenario using open discussion to determine their viability. Do they work? Why? Using the model, dig into the causal relationships. Will the strategies lead to the intended results? Are there unintended consequences?

    One organization, a large technology company, planned to invest heavily in manufacturing an intricate component of its new product rather than outsourcing it. Stress-testing the strategy had an enormous consequence that none of the Team members anticipated – they found that the approach made no sense in any of the scenarios! This powerful message saved the organization a tremendous amount of time, money, frustration and embarrassment. The plan was scrapped, and the manufacturing of the component was outsourced.

Discovering

Executive Briefing: Options

In this session, the Team familiarizes the other executive decision makers who are not on the Team with each of the scenario worlds. The executives are aware of the scenarios, because they have seen drafts of them posted in the Playbook for review and comment. Now the Team vividly brings them to life using whatever presentation techniques best fit the situation – enactments, videos, graphics, artifacts, etc.

The strategy matrix becomes the focal point of the briefing. Each strategy, along with its supporting logic, its competitive consequences and the implications for the organization, is discussed within the context of the various scenarios. The group probes and questions the conclusions represented by the symbols in the cells until they thoroughly understand them. They must not take them lightly or at face value.

The group must take care not to dismiss unpredictable or surprising outcomes of this process just because they may be counterintuitive. The Team may be convinced that a strategy is robust and would have positive results across the scenarios, only to find that the executives identify vulnerabilities that hadn't previously been considered. The Team can add these insights as new intervening variables in the model which may require more rigorous testing of their logic. The results can be quite unexpected, which is one of the benefits of the process.

The Team makes adjustments to the strategy matrix if the Executive Briefing leads to new insights. Once the decision makers are satisfied that the strategy matrix represents the best thinking and information available, they are poised to endorse the strategies they will implement [the most robust] and those that they will keep in reserve as contingencies. Depending on the specifics of the situation, the executive leaders may require more detailed financial analysis or additional input from other disciplines. They may need or want to include Board members prior to making a decision on the strategies. However, it is likely that they will be strong advocates of the strategic recommendations coming out of their conversation with the Team.

Once robust and contingent strategies are adopted, they become the basis for the organization's strategic plan. The Team and the leaders widely communicate the strategies within the organization. The process, reasoning and decisions are fully documented in the Playbook.

The work now moves from strategic planning to strategic action. All parts of the organization will be using the Playbook to develop their more detailed operational plans in the Embodying phase. What has been created to this point is like a fragile sapling, tentatively and urgently seeking sunlight and nutrients. Through Embodying, the branches, twigs, leaves and roots continue developing to support the growth and health of the whole.

Embodying Action

- **Role of Leaders: Integrator**
- **Embedding the Strategy**
  - A Fragmented Process
  - A Whole Systems Approach
- **From Embedding to Embodying**
  - Event Preparation
  - Event Session 1: Appreciating Competencies
  - Event Session 2: Embodying New Approaches
  - Event Follow-up
- **Executive Briefing: Action**

Embodying

Role of Leaders: Integrator

> We are what we repeatedly do.
> Excellence, then, is not an act but a habit.
>
> *Aristotle*
> *384 BC-322 BC*

Many organizations develop a strategic plan only to hit a wall. When strategic direction is "rolled out" within a mechanistic organization, the organization's "immune system" goes on red alert. The initiative is often met with mistrust, resentment and apathy. Each part of the organization sees the strategic direction through its own lens and may misinterpret the strategic intent. When this happens, strategic action is deferred, delayed or subverted. These responses prevent the organization from making any deep or lasting changes. These dynamics can be particularly potent when the strategic direction threatens the entrenched culture, as in the case of becoming a life-sustaining organization.

The Structural Dynamics approach is an antidote to these behaviors. The process itself yields organizational learning – developing shared experience, understanding and knowledge. In the graphic, the diver is coming up from the depths having discovered a holistic way of seeing the organization's situation and bearing a treasure trove of strategic insight. The diver represents the...

 Team Members who arrive at a new level of awareness,

 Leaders who are involved from the beginning,

 Members of the organization who have been tracking and contributing to the initiative.

The insights emerging from this process will light up the organization, illuminating every nook and cranny and all of its internal and external interactions. Insight-based strategy doesn't require internal marketing communications or change advocates to produce results. People naturally see the connections and act upon them; putting these concepts into action is organic.

As an organization becomes life-sustaining, it manifests new awareness, attitudes and behaviors. These shifts occur within individuals, teams, groups, functions, divisions, agencies and the organization as a whole. In nature, nothing has to be told what to do; organisms instinctively know their roles. Through Structural Dynamics, the individual members and parts of an organization move together in a consistent direction with infinite variations on the theme – just as birds form intricate, dynamic patterns as they flock. The collective mind is embodied in each of its components. The organization is experienced through its component parts – each part and each person expresses the whole.

At this point in the initiative, the Team members have become leaders, even those without formal positions of authority. Along with the organization's executive and line leaders, these network leaders take on the role of organizational integrators. Full alignment of executive, line and network leaders focuses the energy of the whole organization. Some leaders participate directly on the Team, and others connect through the influential thinkers they have chosen to serve on their behalf. The Structural Dynamics process fosters a climate of dialogue and collaborative inquiry among all members of the organization.

Embodying

A supportive relationship among leaders of all types is critical to the successful implementation of the initiative; however, a number of powerful organizational forces may work against integrative executive leadership:

- Executives have attained their positions based on their proven expertise. They tend to be strategists by nature, able to deal with complexity and abstraction and to visualize the interplay of a wide range of dynamic forces. It is widely assumed that, as senior leaders, they know what is coming over the horizon. They may find it hard to admit, even to themselves, that the future may be substantially different from the past and that there may be things that have escaped their attention while they've been occupied with their very full agendas.

- Many executives read widely and communicate skillfully in the course of fulfilling their formal responsibility of leading their organization to successfully accomplish its mission. In the Structural Dynamics process, they have enlisted the diverse thinking of the Team to help them achieve that goal. Positional roles do not play a part among Team members. Each member of the Team needs to respect everyone's ideas regardless of hierarchical position. Egalitarianism evolves as the Team coalesces into a high-performance working unit; hierarchical role distinctions fade away. However, if this does not happen naturally, the management of power differentials requires skillful facilitation.

- Senior executives often compete with each other over the organization's strategic direction. Studies of primate behavior [Conniff 2005] show that power struggles between leaders are as natural as alpha monkeys fighting over the warmest rock on the savanna. As primates, we may be hardwired to fight for dominance.[13] Such competition reduces or cancels the impact that leaders could have by acting in concert. Team members, acting with the best of intentions, may find themselves buffeted about by such struggles among their leaders.

- The demands placed on leaders in various organizational functions make it extremely difficult for them to understand the conditions that their peers face [Sales 2008, Oshry 1999]. Misunderstandings lead to mistrust and a lack of empathy for one another, frequently devolving into fierce power struggles. Leaders tend to inflame each others' automatic defensive routines, which then become ingrained patterns in their relationships [Smith 2008]. Team members can get caught in the middle of these conflicts.

These dynamics, inherent in all organizations operating "on automatic," stack the odds against effective leadership behavior. When executives and line leaders are antagonistic, defensive or aggressive with each other, look for these drivers. They are manifestations of fragmented, mechanistic thinking and behavior. Once these forces are set in motion, they interfere with the implementation of strategy. They are the opposite of a whole systems mentality.

Structural Dynamics takes these behaviors off automatic and puts strategy development and implementation into a context of dialogue where people can consider thoughtful, informed ideas together:

---

[13] A vice president of a financial services company [and a former halfback for Army] once contended that no amount of leadership development "charm school" is going to make an iota of difference to the fundamentals of human behavior.

Embodying

- By surfacing environmental forces and weaving them into scenarios, Structural Dynamics drives powerful strategic conversations among leaders that get the organization moving.
- The development and implementation of life-sustaining strategies engages leaders in cross-boundary collaboration by engaging their peers in seeking solutions for the whole.
- By emphasizing the whole system rather than the optimization of the parts, Structural Dynamics makes it clear that "we are all in this together."

When one element of a living organism is performing poorly, the system as a whole is affected: the effect could be some combination of depressed financial results, damage to the reputation of the organization, internal competition and disharmony, or the inability of other parts of the organization to perform to the best of their ability. By elevating this truth into organizational consciousness, Structural Dynamics offers executives an alternative path to conventional organizational mental models and mechanistic approaches to leadership. The life-sustaining initiative creates awareness of the organization as a coherent living system.

Once communication begins to occur more naturally across organizational boundaries, the organization experiences a release of creative energy. People find resources and expertise in places they may not have expected it to be.

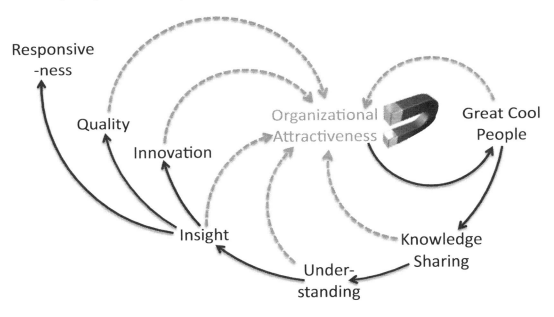

The buzz that develops results in a sharing of knowledge, leading to greater understanding and insight. It can generate a "beehive" of creativity that is highly attractive to other talented people craving this type of work environment [Kunstler 2003]. This hotbed of insights increases innovation, ratchets up quality and improves responsiveness to customer and environmental needs, further reinforcing the organization's magnetic qualities and vitality.

🌱 Embedding the Strategy

To grow and thrive, the seeds of life-sustaining strategies must be sown in the hearts and minds of everyone who is part of the organizational system. How this happens is the tricky bit. We'll look first at the way strategy implementation has been done and is still being done in many organizations. Then we will reflect on the ideas behind a holistic approach. Next we will describe an approach that we have found successful for embodying strategies throughout the organization. This process has its roots in the inclusive nature of Structural Dynamics.

- **A Fragmented Process**

    Most strategic plans fail in the implementation phase. Cândido and Santos [2008] reviewed multiple studies regarding the implementation of strategic planning activities and found that 50% to 90% of all plans fail in the implementation phase. It is such a common problem that the very word "implementation" is a loaded managerial term. Implementation frequently means, "You are to do what we [executive leaders] have come up with and you are going to like it."

    In mechanistic institutions, the strategic planning process requires a great deal of time and effort by the executive committee which, as we have seen, may be dysfunctional and/or consumed with short term issues as a matter of survival. As a result, top-of-the-house strategic plans are produced sporadically, if at all. If there is a top-level strategic plan, it is most likely to be static, i.e., not growing and changing as the organizational and workplace environments change. Because we live in a time of rapid change, traditional strategic plans quickly become irrelevant.

    In large systems where we have asked line leaders, "Does the organization have a strategic plan?" most people respond, "I think so." They assume that someone "up there" must see the whole picture and know where the firm is going. However, it's hard to find anyone who knows what the plan is or how to locate it. The best most people can do is to refer to broad statements in the annual report – statements produced for investor consumption.

    Because organizational strategic plans are not available or, if they exist, are out of date, units, functions, agencies, etc. often produce operational plans in a vacuum. Having nothing with which to align, operational plans at the component level generally seek to optimize the unit's performance, often, unwittingly, at the expense of the whole. Managers base these operational plans solely on:

    ▸ The unit's independent goals and objectives
    ▸ The results they imagine the organization wants from them
    ▸ Executive-mandated performance targets, e.g., 10% expense reduction, 5% sales increase

Embodying

Operational plans feature the well-known "hockey stick"...

"Results were flat or negative this year, but they will turn around next year and continue indefinitely in an upward direction."

– based almost entirely on wishful thinking.

These plans contain a budget estimate of the cost to achieve them.

The budgeting process involves a cumbersome, bureaucratic chain of meetings and reviews, from one level of the hierarchy to another, in assembly-line fashion. Executive leaders approve resource allocations based on the well-scrubbed top-level financial plan. When the operating units receive their allocation, it may not much resemble the budget they submitted. Like a snafu in which a military unit receives a shipment of size twelve boots when no one in the unit wears more than size ten, managers are left to do the best they can with what they have.

Because a unit's operational plan and budget are uncoupled from any organization-wide strategic direction right from the start, the process serves to deepen the estrangement between executives and line leaders. The executives don't understand why the operational units seem to ignore, misinterpret, or misapply their statements regarding the strategic direction of the organization. The units don't know the context in which decisions were made. Individual contributors only understand their own duties; they frequently do not know where they fit into the organization's overall purpose or strategic direction. Everyone keeps their heads down and their fingers pointed at somebody else. With so much disconnection and fragmentation, we do not need to wonder why the overwhelming majority of strategic plans fail.

- **A Whole Systems Approach**
  An organic approach to strategic planning starts and ends with the whole. Each employee and every unit within an organization embodies and reflects the whole organization. If employees don't seem to care or the organizational components act inconsistently, the external perception of the whole is negatively affected.

  The organization's people and its functions are the organization: they are its cells, organs and musculature. They naturally enact the organization's strategic intent within their own domain. When organizations are in alignment with themselves [i.e., acting as living systems], the performance of the whole is optimized. Strategic action at the individual and group level is indistinguishable from strategic action of the whole.

  The idea that each element of the whole embodies the totality reflects the wholeness that we find in nature.

Embodying

For example:
- We see whole vistas even though our eyes are infinitesimally smaller than a scenic panarama, the ocean's horizon or the night sky. We comprehend the totality from our unique, relatively tiny vantage point.
- Holograms contain the whole in every part. When a holographic image is torn into pieces, each element retains the whole image.
- Fractals manifest "self-similar" elements and clusters – think of trees, ferns, clouds, mountains, lightning bolts, snow flakes and broccoli. Every part resembles the whole.

People and components of the organization reflect the whole organization in much the same way. Anyone interacting with an organization forms a mental image of the whole through relatively small touches that leave lasting impressions. If a strategy is embedded in the hearts and minds of everyone in the organization, individuals and groups will embody that strategy in their actions whether they are in immediate contact with each other or not. We know what we are called upon to do within the context of the whole. Living systems rely on the appropriate behavior in all parts of the organization. Attempts to centralize control only interfere with effective action.

The whole is not to be encountered by stepping back to take an overview...
   it is not over and above the parts,
      as if it were some superior all-encompassing entity.
The whole is to be encountered by stepping right into the parts.
A part clearly indicates that the way to the whole is into and through the parts.

Henri Bortoft as quoted by Scharmer, 1999

"Everything is in everything," according to quantum physicist Henri Bortoft [1996]. He advises us not to think of the whole as if it were a separate entity, a thing in itself.

To develop a life-sustaining entity, the operational plans of units need to coincide with the organization's strategic plan – similar expressions of self-similar parts. They are not independent of the whole and cannot exist apart from it. Embedding and maintaining a life-sustaining quality requires every part of the organization to adapt the organizational strategies to the specifics of its distinctive function and conditions.

Embodying

Because every employee at every level is a fractal of the whole, each person represents more than his or her part of the organization. We come to know the organization through its people. For example:

- On a recent trip, our previously positive impression of JetBlue was tarnished by the attitude and behavior of a supervisor who acted more as a hindrance to our travel than as a facilitator of it. We didn't know that JetBlue could be like "that!"
- In contrast, we love shopping at Trader Joe's, where the employees seem "real" – they talk with customers, they know the inventory and, whenever asked about an item, they will drop whatever they are doing and quickly escort the customer to it. Yet, they are not cookie-cutter embodiments of an organizational facade. Every employee's unique personality shines. Compare this with your usual supermarket experience![14]
- An account executive at UBS demonstrates the sort of partnership that his organization offers by calling to discuss some material on our website and how it's helping him think about his own activities.
- The staff of Metaphor Yarns in rural Massachusetts took obvious pleasure in showing us their colorful, variegated stock. They manifested a contagious delight that led us to make a purchase when we thought we were only visiting!
- Our experience with WebEx [now part of Cisco] reminds us what a customer experience can be like – masterful. In fact, we have had the same experience in dealing with any part of Cisco; these repeated experiences have made us into loyal customers.

In a life-sustaining organization, the components and people are in seamless alignment. They are oriented toward common objectives, and they know why they are doing what they do. They have an effective means of feeding information from the periphery back into the organization's learning system – something that is critically important. The organization is not closed off from its environment. It has permeable boundaries; it breathes in everything about its context. People throughout the organization act in concert to create a preferred future.

From Embedding to Embodying

You may be wondering, "How does this happen?" How does each unit and each person in the organization come to embody the strategy so deeply that they effortlessly act in concert? We're off to a good start by creating a cross-functional Team of creative individuals to work on the life-sustaining initiative. As importantly, organizational leaders and Team members have involved the entire workforce throughout the process. We've developed trust by seeking, listening to and considering people's thoughts and ideas. Now, with the strategies selected, we are at a critical juncture that requires everyone to take ownership for what they have participated in creating. It's time to fully embrace the strategies for becoming a life-sustaining organization.

---

[14] "...Trader Joe's mandates that full time workers earn at least their community's median household income, while its 'store captains' can make six-figure salaries...recognize[ing] that better conditions lead to better customer experience and an improved bottom line." [Florida 2010]

Embodying

Although the strategies can be embedded within the organization in a number of ways, we suggest a one- or two-day organization-wide strategic conversation to achieve this goal. We recognize that gathering the whole system in real time and space can be difficult for large organizations. However, if it is feasible, we highly recommend doing so because communications technology has not yet duplicated the multi-sensory experience of meeting in person.

If your organization is too large or too dispersed or both to make such an on-site conversation possible, there are many virtual ways to simulate the same type of interaction. It may not yield the same fidelity as a face-to-face meeting, but by applying creativity, the outcomes of virtual connections can be impressive. One organization held a large event on their "island" in SecondLife.com, a popular virtual reality site on the Internet. The whole organization participated as avatars.[15] Other advancements in communications technology tools are occurring all the time.

EXAMPLE 1: The whole system in the room

A K-12 school system in a mid-sized town decided to hold a two-day meeting shortly before the start of the academic year. Organizers invited all the teachers, administrators and staff, as well as parent and student representatives. They met in the cafeteria of the high school. Classrooms were used for breakout groups. The group spent the first day appreciating their accomplishments and having some "aha" moments as they realized that they did some things very well. Building on this momentum, during the second day, people focused on making commitments for the upcoming year. It was the most cooperative, spirited and academically successful year in their recent history.

Whatever method is used, the objective is to hold a strategic conversation. It is an interactive learning experience, not simply a presentation or an announcement of *fait accompli* strategy decisions to be implemented. This is a dialogic process, with the entire organization participating directly.

The objectives of the event are to:
- Demonstrate the active support and engagement of the leaders
- Engage the entire organization in a review of the strategy work to date
- Appreciate what the organization does well and wants to retain
- Find the best ways to embody a life-sustaining quality throughout the organization

Here is a brief description of the activities that would occur in preparing for, hosting and following up on such a conversation:

- **Preparation**
    1. Identify how everyone in all parts of the organization will participate in the strategic conversation. Along with the groupings created by the organization's formal structure, organizations may want – or need – to mix the groupings during the conversation [e.g., by function, geography, role, etc.]. People can also self-organize into groups at the event using a technique such as Open Space, in which

---

[15] This approach may not be a good solution for everyone because it currently involves a steep learning curve for those unfamiliar with the application.

Embodying

    participants propose a number of discussion topics; they "vote with their feet" by selecting a topic from those proposed [Owen 1997, Vogt 2008]. Whatever methods are used to configure and reconfigure groups, each will have a distinctive voice to add to the strategic discussion.
 2. Create roles for Team members and the organizational leaders. This is an opportunity for leaders throughout the organization to demonstrate their support of the strategic initiative.
 3. Ask the scenario group members to prepare dramatizations of their scenario narrative in short theatrical productions, presentations or videos.
 4. Think through the utilization of electronic communications tools between the various locations to provide live video and audio connections.
 5. Plan to record group activities. The organization will find a wealth of good thinking by mining this data in multiple dimensions. For example, to increase the diversity of the workforce, create a way to collect references for people over seventy, disabled workers and/or gifted young people.

- **Appreciating Competencies**
 6. Ask each group to recall a time when members felt really alive and present at work. Appreciating what is already life-sustaining about the organization and the work environment is an uplifting and inspiring activity that moves people toward a positive view of themselves, their peers and the organization's current capacity to sustain and affirm life. This exercise is based on the principles of Appreciative Inquiry [Cooperrider 2007].
 7. Identify what employees want to hold onto. Don't throw out what is already great. Nurture and grow these qualities.

- **Embedding New Approaches**
 8. Briefly review the scenarios in visual form. Allow time for individual reflection and group discussion of each scenario. Consider the pros and cons that each scenario represents to the organization.
 9. Present the Strategy Matrix and describe the thinking that went into each strategy and how it is seen to work or not work across the scenarios. Take clarifying questions regarding the strategies without debating their merits. Be sure that people understand the logic behind each strategy.
 10. Have each part of the organization engage in a discussion of the robust and contingent strategies and how to apply them.
 11. Check in with the whole system. Not every group needs to report; choose groups whose actions are highly leveraged throughout the organization [e.g., in a technology company, it would be important to hear from engineering group].
 12. Have groups continue working together to decide how they will embody the strategies in their area and how they will measure their own success.
 13. Ask for brief summaries from several groups. Have all groups record their sessions and make them available through the Playbook.

- **Follow-up**
 14. The Playbook contains sections for each group's use. Here the groups detail the action steps they will take to advance the organization's strategic direction: activities

to continue, those to stop and those to start. They also identify how they will measure progress in a way that is appropriate to their work [e.g., customer satisfaction surveys, employee turnover rate, case management, projects completed on time and budget, etc.]. These sections of the Playbook are accessible to everyone within the organization so that groups can learn from one another.

15. The Playbook should allow and encourage creativity. For example, groups could create video reports of this event and follow-up activities as the work progresses. Increasingly, people are relying on visual communication in place of or in addition to the written word.

Consider the following example of an organization-wide strategic conversation.

EXAMPLE 2: Breaking down boundaries in healthcare

A team of research leaders from various parts of a large prestigious healthcare system was interested in finding or developing technologies that would integrate medical disciplines by spanning "technological islands," e.g., the lack of interoperability of devices in operating rooms associated with clinical specialties. The synergies resulting from an integrated record of patient information would improve patient safety and the quality of care. The team addressing this issue discovered several robust strategies. For example, one strategy was to provide non-invasive patient monitoring. This would allow the healthcare system to provide patient care in a wide variety of settings [e.g., hospitals, clinics, homes].

To gain support for these strategies throughout the system, the team convened a strategic conversation. On the first day, the team set up an "Idea Factory" where participants explored and evaluated ideas to achieve technological integration. The purpose of this day was to raise awareness of the power and possibilities of integrated technologies.

The second day, fueled by the energy of the first day, was lightly structured. Clinicians, researchers, administrators and technologists from many parts of the healthcare system participated. The team led the whole group in a discussion about the future of medicine and healthcare delivery – where it might be going and what that might mean for their hospitals. Several high-ranking people vociferously asserted that it would be impossible to break down medicine's traditional discipline-based practices. Others responded in ways that those convinced that change was unlikely were able to hear. Many reasons to support technological integration arose from the discussion. Working in small groups, participants discussed their role in implementing the strategies that the team had discovered. The participants came to realize that the organizational structure and operational practices within the hospital system did not support integrative technologies. For example, researchers throughout the system received grants without a coordinated research agenda. This approach undercut potential advances in patient care and safety. The group concluded that a research center focused on boundary-spanning technologies was required.

A joint effort by these influential people led to the approval of large-scale funding to create a research center dedicated to integrative technologies. This research center continues to make significant contributions to medical science, the invention and manufacture of integrated medical devices and clinical management systems.

Embodying

The previous example demonstrates what can be accomplished. It describes an event in which robust strategies became embedded in actions that affected the entire organization. The following example describes how one group within an organization aligned its activities with the organization's strategic direction. The story demonstrates the results an organization can achieve by engaging its workforce's creativity in customizing strategy to satisfy the needs of each unit and each person.

EXAMPLE 3: Embodying strategies in the workplace

SCAN Health Plan is an HMO Medicare Advantage plan with 110,000 members in California and Arizona. SCAN provides elderly clients with the services they need to live at home independently. In 2008, SCAN developed a 2012 business vision and set a corporate strategic direction. Their objectives included:

→ reducing operational costs
→ breaking down silos
→ becoming less hierarchical
→ being more creative and innovative
→ driving process improvements based on metrics
→ maintaining a high-touch relationship with customers
→ expanding into new markets in a more rapid and cost-efficient way
→ becoming "the employer of choice" – attracting and retaining critical talent, e.g., people with advanced statistical skills

The Workplace Services team, led by Diane Coles, includes members from Business Development, Information Technology, Human Resources, Finance, Real Estate and Facilities working together. With the support of CEO David Schmidt and the executive committee, the team undertook an ambitious multi-year organizational renewal effort, dubbed "Alternate Workplaces Engaging Staff & Office Management Efficiencies," better known as AWESOME.

AWESOME enabled the company to become more life-sustaining by embodying the corporate strategies, including the goal of becoming "the employer of choice." This effort directly tied to SCAN's 2012 vision. The team developed a new approach to workspace planning and provisioning, completely re-envisioning the workplace, supporting telework and reducing corporate real estate costs.

A broad cross section of SCAN's staff took part in space-programming sessions called "design charrettes." In these sessions, employees were asked to think about how they worked and then created a new space concept: townhouses grouped in villages.

Embodying

The following chart, "Aligning Space Design with 2012 Vision," developed by SCAN's Workplace Services team, shows the alignment of corporate strategy and operational activity:

Workplace Strategy
Aligning Space Design with 2012 Vision

Corporate Strategy

Strategy
- Holistic sense of the future
- Member focused products
- Community based resources
- More diverse marketplace

Organization Structure
- Responsive & non-bureaucratic
- Effectiveness culture
- Ad Hocracy
- No Silos

Leadership Style
- Focus on strategy, members, providers
- Innovation & creativity
- SCAN Cares / SCAN Values
- Process improvement / metrics driven

Selection Criteria
- Employee referral program
- Fit with organization
- High performing, independent contributors

Attachment
- Mission driven
- Small company feel / First Call
- 5 Basics
- Senior sensitivity

Workplace Design Strategy

Strategy
- Customer focused design
- Design with future in mind
- Re-engineer process for quicker market expansion

Organization Structure
- New space standards - no longer bureaucratic
- Teaming areas / brainstorming rooms
- Flexible furniture
- Casual areas & café's to break down silos

Leadership Style
- Design to foster creativity and innovation
- Design for process improvement
- Design to increase metrics
- Re-engineer design & furniture process

Selection Criteria
- Flexible workspace
- Remote workforce
- Power of choice

Attachment
- "Branding" in space
- Senior sensitivity training for Architect/Designers
- Integrate member artwork into space
- Call center redesign for better attraction and retention

Embodying

SCAN used electronic Smartboards to design highly flexible team environments within the townhouses. To move to this new way of working, many people would have to accept multiple changes in the workplace arrangements with which they had become familiar.

AWESOME defined five types of work styles that SCAN supports:
- "Fixed address" employees have assigned space.
- "Free address" employees have a shared office.
- "Telecommuters" work from home several days a week.
- "Nomads" work in the office at "touchdown" spaces
-     [a work surface with data and phone connection for nomadic use].
- "Road warriors" rarely come into a corporate office.

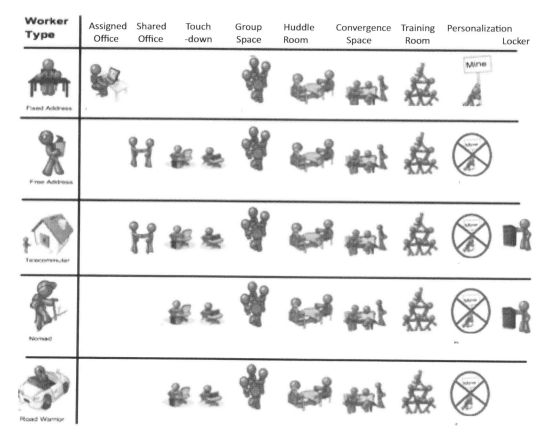

All of these work styles have access to a "village" of group space, huddle rooms, convergence space and teaming rooms. Embodying the new approach led to changes in individual, group and organizational behavior. Diane was SCAN's first Nomad, and she reports that it is "very freeing, very liberating." She finds that she is much more accessible to her staff. The results have been remarkable. Both line leaders and staff members report high levels of job satisfaction.

Embodying

Employees with highly marketable skills, e.g., analytic statisticians, have chosen to remain with the company when they have other options. SCAN's revenues have doubled during a period that some felt it wouldn't survive due to changes in legislation. Overall, SCAN is getting about a 40% payback on its initial investment of $5.7M in AWESOME.

Notably, AWESOME has provided competitive advantages to SCAN's Business Development group – a positive unintended consequence. SCAN can now set up shop in new markets rapidly with a great deal of agility. Workers initially operate from home. Later SCAN establishes a small office to reinforce the culture and to have a presence within the community. Field staff and members of the support workforce at headquarters achieve a deep level of integration through a range of advanced electronic communication systems introduced as part of the AWESOME initiative.

SCAN demonstrates life-sustaining qualities in other ways, too. When the company needed to downsize, executives were honest about what was goingu to happen. The organization notified affected staff a year and a half in advance. SCAN asked them to stay and gain further education. This approach not only helped the employees obtain marketable skills, it kept the workforce intact to maintain a high level of customer service. SCAN's actions communicated a lot about the organization's commitment to its people, both to those who departed and to those who remained.

SCAN's story contains a number of life-positive and business-positive results:

- <u>Creative thinking</u> – SCAN's executive leadership supported innovation by line leaders.
- <u>Fully engaged workforce</u> – Staff members generated options, formulated a plan that they are delighted with and developed their own performance metrics.
- <u>Increased productivity</u> – An overall 18% performance increase by each groups' own metrics. Home workers also eliminated commute time and associated stress.
- <u>Decreased real estate requirements</u> – AWESOME reduced the amount of space required by 25%. SCAN currently occupies 138 square feet per employee, with a target of 122 square feet per employee by 2011.
- <u>Increased ability to attract and retain key staff</u> – SCAN's competitive advantage is enhanced by the company's ability to attract and retain workers with skills that are in high demand, e.g. doctors, nurses, statisticians.
- <u>Business development</u> – Increased agility to enter new markets and initiatives.
- <u>Higher level of employee satisfaction</u>
  - Employees and potential employees are attracted to the option of working from home.
  - The company is able to recruit from a larger catchment area and thus is able to hire the best talent regardless of location.
  - SCAN is able to keep valuable people who move out of the area.
- <u>More appreciation of colleagues in other disciplines</u> – Because they have shared the AWESOME experience and now work together in the same townhouse and/or village, it is now more common to ask how others see things and really listen to their views and considerations. This has resulted in less bickering; one team had their first group hug after a year of working at home.

Embodying

Leaders and Team members continue to be actively involved during this critical Embodying phase of the work. Through Playbook entries and analysis, face-to-face meetings and informal interactions, they stay in touch with people in all parts of the organization. They are particularly attuned to any actions that are not in alignment as a result of misunderstanding or misinterpretation of the strategies. People at all levels across functions will want to talk more about the scenarios, the strategies, and what it means to their groups and to them as individuals to embody a new way of being.

Executive Briefing: Action

The third Executive Briefing takes place toward the completion of the Embodying phase, as the organizational units and functions have begun taking action. The purpose of this meeting is to take a high-level snapshot of the operational plans to see that they are fractals of the organization's strategic objectives [i.e., self-similar parts of the whole]. If questions arise, the unit's leaders will work with the unit to explore what it would mean to embody the strategy. The intent is both to help people think through their plans for alignment and to make adjustments at the organizational level based on new learning. When the executives are satisfied that the operational plans are aligned with the whole, they can be confident that the organization is becoming more life-sustaining. Now, attention shifts to sustaining and expanding the momentum.

Sustaining Results

- **Anticipatory Leadership**
  - Futurist, Strategist and Integrator
  - Balancing Power and Love
- **Why Monitor**
  - Moving Around the Game Board
  - Adapting, Modifying and Turning on a Dime
- **What to Monitor**
  - Signposts
  - Indicators
  - Warnings
- **How to Monitor**
  - Ethnographic Observation
  - Social Network Analysis
  - Productivity Measures
  - Satisfaction Surveys
  - Appreciative Inquiry
  - Strategic Conversations
  - Data Analytics
  - Scanning
- **Sustaining Beyond the Stable State**
  - What happens when we succeed?

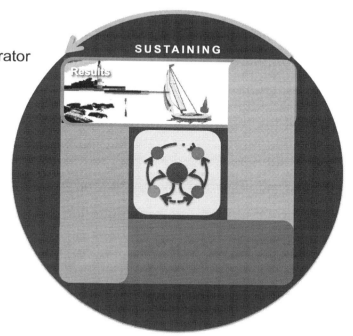

Sustaining

Anticipatory Leadership

I met a traveller from an antique land
Who said: "Two vast and trunkless legs of stone
Stand in the desert. Near them on the sand,
Half sunk, a shattered visage lies, whose frown
And wrinkled lip and sneer of cold command
Tell that its sculptor well those passions read
Which yet survive, stamped on these lifeless things,
The hand that mocked them and the heart that fed.
And on the pedestal these words appear:
`My name is Ozymandias, King of Kings:
Look on my works, ye mighty, and despair!'
Nothing beside remains. Round the decay
Of that colossal wreck, boundless and bare,
The lone and level sands stretch far away."

Ozymandias
Percy Bysshe Shelley, 1818

Very little lasts. Great cultures, including our organizations, are vulnerable and transient. A review of the top companies featured in important management books such as Good to Great and In Search of Excellence reminds us that many seemingly impregnable institutions – Digital Equipment Corporation, Polaroid, Data General, Western Electric, Atari, The Plaza Hotel – can literally evaporate, leaving little behind. And these are the stalwarts, the institutions we thought would always be there.

AT&T was for many years the only telecommunications equipment and service provider in the United States. Known as Ma Bell, it was the premier "widows and orphans stock" – a safe, sure investment. It has been broken up and restructured several times since its glory days. In 2005, a fragment of the former giant that retained the AT&T name was acquired by SBC, itself a spin-off of the original company. AT&T survives in name only because SBC chose to adopt the AT&T brand.

In The Living Company, Arie de Geus tells us that the average life expectancy of a multinational corporation is between 40 and 50 years – far less than the lifespan of most people. He points to several long-lived organizations that he refers to as "living companies" – Shell, Stora Enso, Mitsubishi, DuPont, Alcoa, Proctor and Gamble, and the Hudson Bay Company, for example. A life-sustaining organization isn't necessarily one that never dies; however, it does have certain qualities that reduce its vulnerability and enhance its strength and resilience. It sustains momentum through a conscious awareness of human and natural systems, an openness to organizational learning, and the ability to attract, house and support energetic, creative talent.

It takes a particular kind of leadership to champion and support this integrated approach to strategy development and implementation. We call this way of leading "anticipatory." Anticipatory Leaders can be found in an organization's formal and informal power structures. They are attuned to the world around them, inside and outside the organization. They think holistically; understanding how one event connects with and influences the dynamics surrounding them, they can sense the possibilities for their organizations. They use these skills to dialogue with others, helping them achieve their personal best. The effect resembles Free Jazz.

Ornette Coleman was a master of free jazz, which "features a regular but complex pulse: one drummer plays 'straight' while another plays double-time. The thematic material is a series of brief, dissonant fanfares. As in conventional jazz, each member of the band has a series of solo features, but the other musicians are free to chime in as they wish, producing some extraordinary collective improvisation by the full octet." [Wikipedia] Apart from a predetermined order of featured soloists and several brief transition signals cued by Coleman, the album was created spontaneously, on the spot.

Anticipatory Leaders hear the "music" by tuning in and sensing the harmonics amidst the noise of day-to-day events – even though it can be challenging to find the beat and the melody in times as chaotic as ours. Anticipatory Leaders, like the conductors of great orchestras, are simultaneously in command and deeply respectful of the talent in front of them. They cue the transitions and trust the members of the organizational system to deliver a peak performance. This enables the creative talent within the organization to produce some extraordinary collective improvisation. Anticipatory Leaders let go of the assumption that they know and can control everything and allow themselves to benefit from the wisdom of others. The fresh thinking that emerges can lead to important new insights. These masters of organizational leadership invoke truly elegant solutions.

- **Futurist, Strategist and Integrator**
  Anticipatory Leaders combine the skills of futurist, strategist and integrator. Although we have emphasized different competencies at various phases of the process, these skills occur in concert as a pattern of forces. Although one of the three may be emphasized in a given situation, they are constantly resonating with one another as part of a whole piece. By engaging with others, Anticipatory Leaders teach by example. As a result, these capacities become embedded in people throughout the organization at all levels. Here is a summary of each of these skills:

  1. <u>Integrators collaborate across boundaries</u>
     They convene disparate players to address complex strategic issues, like becoming more life-sustaining. Integrators are skilled at bringing people with differing perspectives into dialogue to elicit creative thinking and insight. They engage all points of view, trusting that better ideas will emerge through fresh, collaborative thinking than could be generated individually. Integrators are able to let go of the myth that they need to have all the answers; they are comfortable asking questions, deeply listening and admitting when they don't know. In doing so, they display intellectual curiosity, emotional intelligence and organizational development skills. [see also "Convening and Embodying – Role of Leaders]

  2. <u>Futurists see systems</u>
     By continually scanning for developments in many disciplines, futurists develop [and encourage others to develop] a big-picture, whole systems perspective. Futurists are inquisitive; they read widely outside their areas of expertise and engage in

Sustaining

conversations with others holding very diverse perspectives. They are able to see patterns where others see chaos. Futurists know that the pressing issue of any particular moment – the "proximal stimulus" – is unlikely to be the only determining factor shaping the organizational environment as the future unfolds. They realize that many interrelated elements influence an organization's ability to be life-sustaining. [See also "Exploring – Role of Leaders]

3. <u>Strategists shape a desired future</u>
   The tension between what is and what could be generates a drive to change. Strategists search for a few principles to guide the development and evolution of the organization to achieve a desired end state. Insight and foresight come from understanding the system of forces driving the present into the future. Strategists rehearse strategic actions to "experience" their effect internally and externally, being particularly sensitive to any unintended consequences. They know that the strategy itself is less important than the result it is intended to achieve. [See also "Discovering – Role of Leaders]

The process we've been laying out, Structural Dynamics, advances the development of Anticipatory Leaders throughout the organization during the life-sustaining initiative. Engaging in a Structural Dynamics process is a powerful way to experience and practice the characteristics of Anticipatory Leadership so that they become ingrained in the organization's culture and norms.

- **Balancing Power and Love**
  To survive and thrive, life-sustaining organizations need to balance their influence in the world [their power dynamic] with their support of emotional coherency [their love dynamic].[16] The power dynamic drives the organization's aspirations to succeed in its mission and to influence others. Without power, an organization is naïve and vulnerable. Love is the recognition that people, inside and outside the organization, strive to express their uniqueness and individuality – in their appearance, their surroundings, their desire to be in community with others, and their knowledge, skills and contribution to the world. Without love, an organization has no purpose.

  Anticipatory Leaders guide their life-sustaining organizations toward a distinctive harmonization of power and love. Every organization has its own balance between the emphasis that it places on power in the world and the emphasis that it places on community. The following chart shows that both are essential to life-sustaining organizations.

  Both power and love contribute to a healthy, functional workplace environment and to the reputation of the organization in the world.

  - The love dynamic creates unity by bringing the energies of the workforce into harmony and connection. Generalists, working across organizational boundaries, help to unify the whole. Team players promote community and cooperation.

  - The power dynamic draws on self-reliance and expertise to move the organization toward its objectives. Specialists contribute a depth of knowledge. Individualists think on their feet and often challenge the organization's assumptions and conventional wisdom, thus mitigating the effects of "group think."

---

[16] This section draws on the thinking of Barry Oshry [2010] and Adam Kahane [2010].

Sustaining

Sometimes these two dynamics are deeply reinforcing; sometimes they exist as tense polarities. Life-sustaining organizations achieve a dynamic balance between power and love that works for them.

Life-Sustaining Balance		Contribution to the organization's environment internally	Contribution to the perception of the organization externally
Dynamic	Role		
Love	Generalist	• Builder of culture and community • Promotes shared language, knowledge, process skills and capacity	• Parts seen as supporting each other in the service of the whole • Strengthens organizational brand and identity
Love	Team Player	• Supporter of peer relationships • Champion of a common purpose • Facilitator of harmony and cooperation	• Experienced as a coordinated system, one that "has its act together" • Responds with integrity to all its stakeholders • Steward of the environment
Power	Specialist	• Adds a richness and variety of ideas and experiences • Deepens systemic capacity differentiated by function and role	• Deep and broad expertise • Acts quickly and appropriately to threats and opportunities • Responsive to customers
Power	Individualist	• Tests, challenges and stretches the system	• Resilient, flexible • Able to adjust course on the fly • Responsibility resides at all levels of the system

Anticipatory leaders engage the workforce in the awareness of the organization as a whole, creating a sense of community and strength. They foster a work environment that encourages everyone to realize their full potential, knowing that, by doing so, individuals will not only fulfill themselves, but they will also strengthen the power of the organization. Making the relationship between power and love conscious sustains the capacity of the organization to derive full benefit from the life-sustaining strategic initiative. In the Sustaining phase, Anticipatory Leaders at all levels of the organization engage in strategic conversations with their reports, peers and leaders to be sure that both power and love dynamics are present in a balanced way. These conversations keep awareness focused on maintaining the health of a life-sustaining organization.

Sustaining

> **EXAMPLE: Anticipatory Leadership**
>
> Angelo Lamola is an Anticipatory Leader. As an executive at Rohm and Haas Corporation, he sponsored life-sustaining initiatives to harmonize the power and love dynamics in his organization. A cross section of line leaders and technical and administrative staff members across the organization designed a series of "Community Days." Angelo opened one by saying, "I encourage each of you to publish your resumés broadly. I know you can get other jobs easily. I want you to freely choose to be a member of this organization every single day you work here. I want you to know that this organization wants people working here that every other organization wants to have working for them. I want this day to embody why you've made the choice to be part of this organization rather than another one."

Why Monitor

A life-sustaining organization is one that learns to recognize and reinforce its strengths as it evolves new competencies. It continuously reflects on changes in its environment as it simultaneously pursues its objectives. It grows in its understanding of future objectives while being responsive to current reality.

The diagram shows that innovation, quality and responsiveness – derived from being life-sustaining – lead to positive results for the organization.

Life-Sustaining Organizations Are Magnetic

Sustaining

People are attracted to winners! The idea of what is known as the "success to the successful" dynamic [Senge 1990]" is that those who are successful tend to be at the head of the line when the next opportunity comes along, even if they are not innately superior to their peers or competitors in any fundamentally important way. They advance while others stagnate – with unintended consequences for individuals and their organizations. A successful organization becomes more attractive to the talent it needs. It is the employer of choice for top candidates because it is a winner, has many talented people and is engaged in projects that make a positive difference in the world. It expresses both power and love in the workplace and with regard to nature. Multiple reinforcing loops create a virtually unbreakable cycle of success.

- **Moving Around the Game Board**
  Through disciplined scanning of the external environment, the organization will continually monitor how the future is emerging and which scenario it most closely resembles. The future won't unfold exactly like any of the scenarios; it is always full of surprises. The world may look as though it is heading toward a Collapse scenario, only to have a major new technology emerge that shifts the action toward Discipline or even New Reality. How the world looks can vary based on your vantage point. The poor are likely to see developments in politics and economics much differently than those enjoying lavish lifestyles. Our view of the future also depends on when we look. Immediately after a major terrorist attack or catastrophic natural disaster, a Collapse scenario will seem more relevant. Similarly, a major technology or scientific breakthrough may give us an excessively rosy outlook. Multiple scenarios could play out simultaneously in various locations. Or very different scenarios could occur sequentially. By continuously monitoring external events, measuring internal responses and results, and revisiting assumptions, the organization drives agility deeply and broadly within the organization.

- **Adapting, Modifying and Turning on a Dime**
  A contingent strategy might be viable only in one or two scenarios; for example, a strategy to use holographic images for enhanced communications between remote colleagues may be contingent on the emergence of a Discipline or New Reality scenario. Monitoring what is happening is key to knowing if a contingent strategy should be implemented and when to do so.

  "Ride Out and Maybe"[17] is a great way to think of embodying life-sustaining strategies: their success is based on "maybes." Nothing is guaranteed. Only a really sharp, committed workforce can respond effectively to the multitude of uncertainties in our complicated world. This is the clear rationale for developing life-sustaining qualities that attract and hold high-caliber people: they are able to deal creatively with complexity and change. When a discontinuity occurs, they are prepared to act quickly because the organization has prepared contingent strategic options. These workers are off and running, taking advantage of opportunities as they emerge and avoiding threats while others are stymied by unexpected events.

---

[17] Bob Dylan may have had a publishing company by the name "Ride Out and Maybe."

Sustaining

The life-sustaining organization uses new information and understanding to establish the next cycle of strategic thought and action. Cycling through the Structural Dynamics process deepens strategic thinking, planning and action skills throughout the organization. Rapidly iterating the process builds capacity within the organization. Each iteration increases the organization's nimbleness – its ability to learn, change and grow. The learning process is never complete. Through continuous practice and expansion of its field of vision and imagination, the organization develops "strategic intuition" [Duggan 2007]. Its reflexes become finely honed. Every few years, it will want to take a fresh look at its life-sustaining strategies, reexamining them in some depth. It may have to do so more often if there are dramatic changes occurring in the organizational environment.

What to Monitor

A life-sustaining organization monitors the impact of its actions both internally and on the external environment. Fortunately, the scenarios provide a strong early detection system. When the stakes are high, knowing what to look for is critical. Through the scenario work, the organization knows the implications of various signposts, indicators and warnings; a number of possible futures have been anticipated and strategic responses have been rehearsed.

- **Signposts** mark <u>changes within the organization</u> – moving toward or away from intended outcomes. They act as guides when dealing with complex situations. A lighthouse definitively marks a position; to mariners in stormy, perilous situations, it is a guide to a safe harbor. A signpost guides the organization toward a desired result through the causal interaction of internal factors and forces. For example, employee satisfaction is a signpost that indicates the results of many interacting influences.

- **Indicators** reveal the <u>impact of changes internally and externally</u>. A series of red "nuns" and green "cans" guide ships along deep channels, indicating where it is safe to navigate. For example, high employee satisfaction might lead to more applications for each open position. It could also lead competitors to raise the bar to attract creative talent. Consider what indicators are critical to track and how often.

- **Warnings** are <u>events in the external environment</u> that provide information regarding an impending change toward a particular future condition – like a marker that shows the location of obstacles just below the surface in an otherwise navigable area, perhaps requiring a change of course. For example, a major breakthrough in communications technology that makes virtual meetings a more seamless and natural experience might indicate the possibility of increased remote work, outsourcing or off-shoring.

These distinctions – between signposts, indicators and warnings – are helpful in thinking about the wide range of variables to be watched. Once the organization has identified the factors to monitor, it tracks and documents them in the Playbook. The information is thus widely available to those best able to grasp the significance of changes in these areas.

Sustaining

How to Monitor

Many tools exist for assessing the impacts and results of strategic actions. Here are a few that we have used to good effect:

- **Ethnographic Observation**
  Ethnography can be used for close observation of the workplace environment. It is an established field with many subdivisions, including organizational ethnography as practiced by John van Maanen [1988], Terry Deal [2000, 2008] and others. Ethnography examines everyday behavior and explores the drivers, values and thinking behind it. Anika participated in a research project at Bell Labs that developed an observational approach called "rapid ethnography" [Millen 2000]. The core elements of rapid ethnography include identifying the research scope; interviewing key players; capturing rich field data with cameras, recordings and notes; interactive observation techniques; and collaborative systemic data analysis. Rapid ethnography includes the people being observed as participants in the process of discovering what is actually occurring, rather than what we expect or assume is happening [Norman 1998]. For example, if an organizational norm emphasized peer-to-peer responsiveness, rapid ethnography could be used to determine if that behavior was actually occurring.

> **EXAMPLE: Observation in Action**
>
> When a consulting firm landed a large contract, it needed to rapidly double in size from 15 to 30 professionals. The original crew knew that how it expanded and whom it chose to join the small firm would be critical to its future. A few people who didn't fit could negatively impact the organization's life-sustaining culture. The 15 observed that, even though they had a pretty high level of conceptual diversity [i.e., they frequently disagreed over facts and conclusions], they all worked well together. They decided to look for the "glue" that made them a high-powered team. Using rapid ethnography, the observers noted a common trait among this set of individuals. They each demonstrated a nimbleness in being staunch defenders of a position one minute and able to defend the opposite position in a flash. Digging deeper, the researchers discovered that most had some experience on debate teams either in high school or in college or both. This insight gave the firm a reference point for looking at prospective recruits. It has since grown quite large and has become one of the dominant players in consulting internationally.

- **Social Network Analysis**
  Analysis of the social ties that people inside the organization have to each other and with people outside the organization can reveal valuable information. In a memorable *Harvard Business Review* article, "Competent Jerks, Lovable Fools, and the Formation of Social Networks" [Casciaro and Lobo 2005], the authors term warm, trusted network and line leaders "lovable fools." It may be hard to point to any specific contribution they make, while "competent jerks" can demonstrate their worth by possessing specialized competence. The fools tend to get laid off first in a downturn, often with devastating consequences. It turns out that the so-called fools are frequently the hubs of social networks; they focus on the love dynamic, keeping people in communication with each other, building community and making the work experience more effective and

Sustaining

rewarding. Disrupting a social networking node can risk a plunge in morale and bottom-line results.

As discussed in Convening and Embodying,[18] tight connections between leaders are rarer than we would like them to be. The functional strains and geographic dispersion of complex systems lead to alienated relationships among them rather than teamwork. When leaders do have strong positive ties, they start collaborating across boundaries, and the organization experiences a wide range of benefits. Even though organizational systems running on automatic create alienation between leaders, conscious living systems need to connect these hubs of the social network, ensuring that the power dynamic is in balance with the community aspect of the love dynamic.

While organizations can use social network analysis to monitor how well strategies are being embodied, the way they use this approach must be consistent with the idea of being life-sustaining. Covert monitoring of employees, which seems to be on the rise,[19] would be the antithesis of life-sustaining behavior.

Controversy swirling around the use of email, Internet surfing and social media sites [e.g., Facebook and LinkedIn] in the workplace may be an indicator of the degree to which an organization has become life-sustaining. When the energy of employees is at odds with the aspirations of the organization, these resources can be used as a diversion and therefore be seen as a threat. When the energy of the workforce is aligned with that of the organization, the same tools can support productivity and enhance the experience of working in the organization.

- **Productivity Measures**

  Measuring results is a powerful demonstration of the value of being life-sustaining. Seeing what's been achieved on a consistent basis makes a clear statement about an organization's priorities and values. It is said, "You get what you measure." An organization may find it convenient for a call center to measure calls per minute, for example. But if the quality of the customer experience and the outcome of the calls aren't taken into account, it can be a destructive measure. There is no one correct or universal measure of productivity. Each organization and each part of the organization must determine which measures reflect the results that they want to achieve. Life-sustaining organizations might measure [Senge 2008]:

  ▸ *Pollution prevention, use of clean technology and sustainability practices*: the results of efforts to reduce their carbon footprint by minimizing consumption; reducing waste and emissions from operations; using recycled materials; designing products for reuse and recycling; using nonpolluting, renewable fuel sources, etc.

  ▸ *Community responsibility:* the degree to which an organization has increased its involvement with stakeholders to maintain and improve its reputation and legitimacy through transparency and connectivity.

---

[18] See "Role of Leaders: Integrators." In Convening, we discuss the connections between line leaders [pages 21 – 23]; in Embodying, between executive leaders [pages 93 – 94].

[19] SpectorSoft, a company that documents and archives employee computer and Internet activity, has been named one of the fastest-growing private companies in America by *Inc.* magazine for five years.

- **Satisfaction Surveys**
  As an indication that employees are embracing and sustaining the strategy, both customer and employee satisfaction surveys should show high scores and steady improvement. However, organizations must exercise a great deal of care in using surveys. In our experience, many surveys measuring satisfaction levels lack the necessary subtlety and nuance to discover the outcomes that need to be tracked.

> **EXAMPLE: Are you satisfied?**
>
> A personnel manager at one of the "100 Best Small Business Places to Work" was featured in a conference-call seminar to share the company's best practices with other companies hoping to attain a place on the list. This small business had been growing rapidly and winning recognition for the quality of its work environment and the motivation of its workforce. The manager emphasized how much employees enjoyed the company's regular pep rallies and how enthusiastically they participated. Asked, "What if someone doesn't really get a charge out of pep rallies?" she responded that the company has ways of finding out who doesn't want to be part of its high-performance culture. Every employee is evaluated regularly on a number of measures, including cultural fit. Each year, those ranking in the bottom 10% are laid off!
>
> In our opinion, this company is missing important elements to sustain the quality of life in its workplace. We suspect that a good number of employees "game" satisfaction surveys in companies like this one. When Deming [2000] wrote about "driving out fear," he was focusing attention on organizations as systems, systems capable of learning rather than finding fault and blaming individuals.

The whole notion of "satisfaction" may need to be revised. "Joy" might be a better measure for life-sustaining organizations. Instead of asking, "Has management created an open and comfortable work environment?"[20], life-sustaining organizations might ask, "Are there instances when you felt absolutely ecstatic about your work environment? If so, can you describe a few?"

- **Appreciative Inquiry**
  Rather than "search for the guilty," Appreciative Inquiry is a quest for the great. This approach looks at the positive attributes of individuals, teams and organizations. By appreciating what they do well and noticing when they succeed, organizations can seek to learn from their strengths and adapt those assets to emerging situations. Emphasizing and celebrating what's working results in a different type of inquiry and learning that shifts the focus from what's wrong to what's right. Rather than fixing issues and solving problems, organizations grow and expand their healthiest and most beneficial aspects. This is the difference between focusing on wellness instead of pathology. See the Appreciative Inquiry Commons[21] at for more information on this life-sustaining approach to organizational development.

---

[20] as asked in SurveyShare.com 2009

[21] www.appreciativeinquiry.case.edu

Sustaining

- **Strategic Conversations**

  Analyzing a few strategic conversations can reveal volumes about the degree to which the organization is implementing a life-sustaining strategy. What are people talking about? If there is a strategic intent to build a more diverse workforce, how does that show up in a product launch discussion or in the analysis of market research statistics? If there is a desire to reduce the organization's carbon footprint, do gripes about commuting hassles gets transformed into transportation options? It is a good idea to spend a few minutes at the end of every formally convened conversation to ask, "What did we do and how did we do?"

  Every part of an organizational system has a theory for being effective that can be inferred by observing its behavior rather than from what it says or believes about itself. Chris Argyris and Don Schön [1995, 1978] created a number of methods for recording "directly observable data" as a way to explore the match between "espoused theory" and "theory-in-use." An espoused theory is an assertion of what people want to accomplish through their actions, e.g., "Encourage more candor in the workplace to enable discussion of critical issues." A theory-in-use is the set of related hypotheses people hold as inferred from their actions, e.g., "Keep the work environment pleasant by discouraging criticism." Even a small amount of data from important conversations can yield a great deal of insight. For example, explicitly inquiring about points of disagreement or ambiguity is a good way to determine the degree of openness an organization exhibits under pressure – as contrasted with how open it believes it is [Sales 1984].

  Consider if the organization's behavior is consistent with its espoused principals and beliefs. Analyzing what is going on in strategic conversations is a good way of deciding on the type of training most needed to help an organization become more life-sustaining. Take notes or record some of these interactions to be able to engage in more detailed reflection.

- **Data Analytics**

  Organizations use data mining and analytics to understand all sorts of activities. Retailers apply these techniques to anticipate consumer buying patterns [Berson, Thearling and Smith 2002]. Amazon, Pandora and Netflix are using data mining and data analytics when they suggest books, music and videos you would enjoy [Davenport and Harris 2007]. Google's PageRank algorithm presents search choices to users extremely rapidly by analyzing complex relationships between nodes found in a huge volume of data. This technology is being adapted to understand many types of networks [Burns 2009]. A large professional association has been deepening its understanding of the latent service needs of its corporate members by doing a closer analysis of teleconferences and in-person meetings. Life-sustaining organizations probe documents and public conversations for insights regarding what is of interest to the workforce for everything from food choices in the cafeteria to partnering with Barnes and Noble to provide effective work environments for the remote staff.

  Technological breakthroughs are poised to create radical new vistas for data mining. Microsoft is experimenting with "total recall," expanding technologically assisted e-memory to the point that virtually all actions can be captured and analyzed according to an emerging set of criteria [Bell and Gemmell, 2009]. The capacity required to store and analyze this amount of data is daunting.

Sustaining

Imagine what it might mean for workplace design to engage the workforce effectively using data analytics. *Excellence by Design* [Horgen, Joroff, Porter and Schön 1998] looks at success and failure in workplace design processes. The authors found that collaborations engaging end users had a much higher success rate than any solutions imposed by architects working solely with executive decision makers. Data analytics promise to make the capturing and interpretation of information from occupants of the work environment much easier and, by so doing, reduce the risk associated with large capital investments.

- **Scanning**

  Scanning the external environment is important throughout the Structural Dynamics process. In Exploring, the Team scans for events relevant to the Decision Issue. In Sustaining, scanning keeps the organization on its learning edge. The Team needs to stay abreast of all developments regarding the critical uncertainties that it has identified. The STEEEPA categorization [Societal, Technical, Economic, Educational, Environmental, Political, Aesthetic] suggested in Exploring can be used here as well.

  The following "World System"[22] schema has evolved over time – as living systems do. We include it to illustrate another way of grouping events to monitor.

World System Scanning Categories		
Global Governance: international bodies, demographics, human rights, world futures, green policy	Science/Technology: scientific frontiers, new technologies, outer space, cyberspace, communications	Global Economy: finance, trade, taxation, poverty and welfare, business, economic thinking, economic crisis
Government: renewing government, improving democracy, deficits/ spending	Environment/Resources: biodiversity, waste/pollution, climate change, sustainability	Development: antipoverty, vulnerable regions, technology transfer, land use
Society: race/gender/ ethnicity, religion, lifestyle, culture, families, sports	Health and Health Policy: healthcare reform, disease, new drugs and medtech	Energy: overviews, oil and gas, nuclear, renewables, conservation
Food/Agriculture: food issues, genetic modification, farming, organics, hunger	Work and Organizations: labor issues, incomes, new jobs, unemployment	Security: terrorism, weapons, defense spending, peacemaking
Regions and Nations: Americas, Europe, Asia, Africa, Middle East	Education: school reform, higher education, adult education, learning	Crime/Justice: violence, corruption, prevention, police, courts, punishment
Cities: new urban visions, planning, greening	Transportation: low-carbon and no carbon mobility	Business: green businesses and economics

---

[22] This schema was adapted from the work of Michael Marien [1979-2008 and 2010].

Sustaining

Whatever approach you use, including one of your own invention, we encourage organizations to continuously, vigorously and broadly explore events that may be relevant to their unfolding future. Having done structural modeling and scenario analysis, the organization knows the causes of relevant events and their effects. By tracing the impact of events in multiple domains through the structural model, the organization can better grasp their implications and rigorously consider what is going on. Such events no longer seem isolated and surprising. Continuous scanning sustains the anticipatory sensibility an organization develops during the life-sustaining initiative. Scanning is mentally expansive, contributes to an organization's sense of place in its environment, monitors the direction in which the organization is moving and...it's fun!

Sustaining Beyond the Stable State

Many indicators point to a future rife with discontinuities. Scientific and technological forces are driving change; many economies, societies, institutions and ecological habitats are fragile. Everything is interconnected. A moment's notice may be all that's available in an era of perpetual change. All of us are going to be caught by surprise, and many of our organizations will respond sluggishly to seemingly "out-of-the blue" developments, not knowing how to respond. With such powerful chaotic forces in motion in so many domains of human existence, many unprepared organizations will not survive.

Those who have developed well-honed early warning systems are in a much better position to sense the "tsunamis" and to take action. Structural Dynamics is designed to create the conditions in which organizations see a broad landscape of possibilities so that they are not completely astonished and overwhelmed by events. They may not know every possibility or every detail; however, having considered a full set of scenarios, they will have the skills to process new information quickly. Continuous sensing, scanning and analyzing are key to sustaining this vitally important organizational learning system.

- **What happens when we succeed?**
  "Success to the successful" reflects the workings of "power laws," where a few successful nodes in a network have tremendous leverage over an entire system [Strogatz 2003]. The effects can be dangerous if the power is in the hands of despots. On the other hand, success often reflects a positive evolution toward a new and better way of doing something. We learn each time we effectively intervene in the workings of a system. If we are able to apply that learning to living systems that are constantly evolving, presenting us with new situations and new challenges, we progress in our understanding of the system's structure. We become better able to influence its behavior. In this way, we can maintain the life-sustaining qualities of our organizations!

  Companies like Google, Apple and Facebook have reshaped the world of work. The companies on the Fortune list of Best Companies shape their industries. On taking a close look at any of these firms, we could probably find many aspects to criticize. However, it is highly instructive to see the various ways in which each of them manifests life-sustaining aspects. While perhaps not perfect, these organizations have a worldwide impact. We could say the same about Ivy League universities in education or the leading teaching hospitals in the healthcare domain.

Sustaining

The organizations that sustain highly distinctive, life-positive cultures are the stars, emulated by hundreds and thousands of others. The adulation they receive has a tremendous effect on their internal dynamics: they are able to recruit great people because everyone wants to join them. Their success sustains their efforts. These reinforcing dynamics create a life-sustaining organization's magnetic field.

Part 2: A Case Study

Introducing GoGo Global Transport
- GoGo Past and Present
- GoGo's Future

Convening the Team
- Role of Leaders: Integrator
- Preparing for the Life-Sustaining Initiative
- Convening

Exploring Facts
- Role of Leaders: Futurist
- Scanning Events
- Recognizing Patterns and Structure
- The Heart of Structural Dynamics
- Exploring Deeper
- Executive Briefing: Facts

Discovering Options
- Role of Leaders: Strategist
- Articulating Scenarios
- Identifying Options
- Executive Briefing: Options

Embodying Action
- Role of Leaders: Integrator
- Inspiring Commitment
- Executive Briefing: Action

Sustaining Results
- Anticipatory Leadership
- Executive Briefing: Results

We present the GoGo case to illustrate the application of Structural Dynamics to becoming a life-sustaining organization. It is a composite case based on several engagements with organizations large and small, in the private and public sectors. We provide the level of detail necessary for the reader to understand the use of the methodology. We do not supply a complete set of scenarios or other elements of the process because they are unique to each situation. We don't want to give the impression that these solutions are directly applicable to your situation.

GoGo Case

Introducing GoGo Global Transport

GoGo Past and Present

GoGo Global Transport designs, manufactures and services electric vehicles used by industry [e.g. fork lifts], government agencies [e.g. utility vehicles] and recreational facilities [e.g. golf carts]. GoGo was founded in 1948 in St. Louis, Missouri by Herbert Chaffrey. His family had done well in the auto-supply industry. Herbert grew up in the country-club set. His imagination was sparked by an early version of the golf buggy. He improved the design and began producing them in a rented warehouse.

In 1965, GoGo acquired a small fork-lift manufacturer as a natural extension of its line. In 1990, GoGo's R&D department developed a line of lightweight, battery-powered utility vehicles that are now widely used in park systems and municipalities throughout the United States.

Herbert's son, Prescott, a Babson MBA, took over as president and CEO in 1998 when Herbert retired at 74. Prescott's sister, Denise Chaffrey, holds a degree in Operations Management from the University of Michigan and is the senior vice president of Manufacturing.

Prescott Chaffrey NYC 2009

Other members of GoGo's executive committee [known internally as as the GGEC] are:
- John Dougherty, Chief Financial Officer, MBA, Washington University
- Sheila Hardy, Chief Operating Officer, M. Eng., University of Indiana
- George Volt, SVP Sales & Marketing, MBA, University of Southern Illinois
- Tom Kennedy, SVP Human Resources, MA, American History, Vanderbilt
- Bud Schiller, Chief Information Officer, MS, Drexel University
- Andrea Stein, SVP Engineering, M. Eng., Rensselaer Polytechnic Institute

GoGo has been trading on NASDAQ since 1987, but it is still family controlled. Between 2000 and 2007, the organization averaged growth year to year of:
- 6% revenue
- 8% profit
- 3% in number of employees

In 2007, its revenues reached $2.7 billion, and it employed 14,000 people. In the fall of 2008, the financial markets went into decline, leading to what the CEO calls the "Great Disruption." GoGo's golf cart operation was immediately hit hard. Credit availability became extremely tight, introducing serious constraints on all of GoGo's expansion plans. Growth in revenues and profits was flat to slightly negative throughout 2009. In the larger context,

GoGo Case

U.S. unemployment hit 10.0% in December 2009.

The company is proud of its heritage as a family business with strong ethical traditions that are expressed in its Mission and Values statements; these were refined in 2001 and are posted on placards in all GoGo's conference rooms.

> **GoGo's Mission**
>
> GoGo is committed to building the world's finest electric vehicles.
> We price our products to attract customers away from carbon fuel alternatives.
> We are committed to the health of the planet.

> **GoGo's Values**
> - We treat our people, our suppliers and our customers with respect, honesty and integrity at all times.
> - We believe that employees who are treated well and given challenging work will achieve their personal best.
> - We believe the attention we pay to our employees will be reflected in the way they treat our customers.
> - We believe that both employee satisfaction and customer satisfaction are directly related to our ability to fulfill our mission.
> - We care about our planet and take care of the environment in everything we do and every product we produce.

- **GoGo's Facilities**
  Administration and Engineering: GoGo is headquartered in a historic office building in downtown St. Louis.[23] The company also has several administrative spaces which are leased in facilities nearby. GoGo is negotiating to buy a second office facility next to its original HQ tower to eliminate the need for these spaces. The company also has leased sales and service offices worldwide. Fifteen percent of GoGo's office staff work from home and have no assigned workstation in a GoGo facility. Even some executives have chosen to work remotely and give up their offices. GoGo provides a small stipend to help defray the costs of setting up a home office. The offices and cubicles that would have been assigned to these employees are held in reserve to allow for growth, moves, drop-in occupancy and other miscellaneous uses. The other 85% of GoGo's workspace is assigned to individual employees. However, at any given time, only about 35% of the offices and workstations are in use.[24] Office and workstations are a standard size across all functions and business units.

---

[23] Go-Go bought the building in 1993 after leasing space in it for many years.
[24] Employees are often traveling, working off-site, in conference, at lunch, out sick, on vacation, etc.

GoGo Case

Manufacturing: The company owns three manufacturing facilities in the U.S., one in Mexico and one in Belarus. GoGo is in the process of contracting with a manufacturer in China to produce some its vehicles. The company had budgeted for significant investments in IT to keep its factories state-of-the-art and support remote employees before the economic downturn.

In January 2010, the lingering financial uncertainty in the United States and elsewhere is causing growth in GoGo's traditional markets to stall or decline. GoGo's fork-lift business is now dominated by larger competitors. The golf cart business is directly tied to economic conditions. On the upside, Prescott and the GGEC see new opportunities emerging for GoGo's skills and expertise. The military has a growing demand for all-terrain utility vehicles. Japanese and European auto manufacturers are creating a worldwide market for electric vehicles: a U.S. company has begun producing a medium-speed electric vehicle [MSEV] after successfully piloting a neighborhood electric vehicle [NEV].

Competitor's NEV

GoGo executives realize that their employees have a lot of energy and enthusiasm for venturing into such emerging markets. They also know that GoGo would have to be able to attract some of the most creative talent coming out of the best engineering, technology, business, design and science programs to become a viable player in these new fields. If they succeed in hiring these highly sought-after people, GoGo would have to feel assured that the company could keep them challenged, satisfied and engaged so that they would not be attracted away by competitors. The GGEC is quite aware that the future is shaping up to be a complex playing field for GoGo. They have decided that 2010 will be the year that GoGo takes a hard look at its strategic options to ensure that the company is ready to take advantage of opportunities and defend against growing threats.

GoGo's Future

January 4, 2010: GoGo's Executive Committee [the GGEC] meets every Monday morning.[25] The GGEC has dedicated its first meeting of the year to an open exploration of what GGEC members are thinking and concerned about. Much of the conversation focuses on how to do more or the same with fewer people:
- Tom, S.V.P. of H.R., notes that simply filling vacant positions will not move the company in the direction it wants to go. He outlines the skills GoGo needs to expand and enter into new product lines such as neighborhood electric vehicles.
- Bud, C.I.O., still smarting from recent budget cuts, points out that GoGo will steadily be put at a competitive disadvantage if current trends continue; the information technology specialists GoGo most needs will be attracted to companies who offer the best tools for research and development.
- Sheila, C.O.O., whose responsibilities include real estate and facilities, adds that the company has an opportunity to save money *and* get better tools if it adopts a policy strongly encouraging employees to work from home. GoGo would then be able to

---

[25] The preference is for executives to attend in person. Those who are traveling or located outside of St. Louis call in.

reduce the amount of physical space it requires and could deploy the savings to technology resources.
- Tom immediately and vehemently objects to the "simplistic" idea that such functions as Research & Development could be conducted without having a place to work together as a team.
- Andrea, Engineering S.V.P., who has R&D in her domain, seems to be moved by Sheila's argument, but she says she needs to consider the pros and cons of the idea because she has no experience in managing a distributed organization in an arena like new product development. She does agree, though, that the mix of skills in her department needs to change to keep up with the times.
- Denise, Manufacturing S.V.P., points out that, because GoGo has international facilities and markets, the company has to improve its ability to embrace people from a wide variety of cultural backgrounds.
- George, Sales and Marketing S.V.P., agrees with Denise and adds that he has found that building a diverse workforce is easier said than done for an American company with strong Midwestern roots.
- John, CFO, says that, with U.S. deficits and various sorts of instability in China and India, he does not see many reasons to become concerned about GoGo's ability to hire people well into the future. As far as he can see, the labor glut looks like a permanent fact of life.
- Andrea notes that governmental regulations regarding the environment, energy usage and product efficiency will require GoGo to act quickly upon some of its new thinking such as fuel-cell powered equipment and to get moving in new directions that they haven't yet imagined.

Prescott appreciates the conversation and the importance of the issues being raised. He understands that none of them can be resolved independently of the others. He says he would like the GGEC to engage GoGo's workforce in a dialogue regarding the organization's ability to face the future with confidence. In his words:

> "We all know something about how to recruit people and motivate our workforce. Let's look at the human side of GoGo by creating a process where we gain insight from some of our most strategic thinkers and use that learning to move forward in sync with one another in everything we do. I am willing to invest GoGo's time and resources in this effort because I expect it will deliver key insights into these challenges."

Prescott asks Sheila to take the lead in pursuing this initiative. Sheila is familiar with Structural Dynamics as an approach to strategy planning, and she proposes to use this process to guide the work. With the agreement of the GGEC to move forward, Sheila asks her colleagues to help identify big-picture thinkers in their areas to work with her as members of a Team. She emphasizes that she wants the whole system represented; all levels, departments, ages, genders, races, ethnicities, etc. that exist within GoGo.

Prescott stresses that he wants the executives to stay connected throughout the process and to participate in briefing sessions at key points, particularly when the Team has established the facts, when they have developed options, in reviewing the actions taken throughout the organization and in assessing results.

GoGo Case

Convening the Team

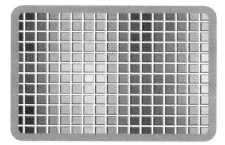

🌐 Role of Leaders: Integrator

Sheila knows that the initiative must be an open process, with transparency throughout – an open flow of communication in all directions, with all parts of the organization and with every employee. Prescott, Sheila and the other executive leaders play critical roles; they will be building a Team of diverse thinkers chosen for their strong, often contrarian, views. Integrative leadership is required early in the initiative.

Sheila will be carefully promoting a widening of possibility rather than a hardening of positions and wants a facilitator who will be her strong ally in this effort.. She is aware that while Team members may not be highly skilled in listening to others and passions may be inflamed by disagreements, the process is designed to be inclusive and expand thinking. A wide range of perspectives sparks divergent scenarios,

The leadership skills of Team members will be developed throughout the process. The GGEC is excited by the opportunities for their own leadership development as well as the opportunity to provide for the development their staff members participating on the Team.

🌐 Preparing for the Life-Sustaining Initiative

In preparation for the initiative, several tasks must be completed such as choosing a facilitator, forming the Team, setting up the Playbook and securing space.

- **Choosing a Facilitator**
  So that she can fully participate in the strategic initiative, Sheila asks Donna Smith, one of GoGo's organizational development professionals, to facilitate. Donna has well-honed facilitation skills, and she is widely respected and trusted within the organization. Donna has attended Art of the Future's trainers workshop and has shared a summary of her learning on GoGo's intranet site [which is why Sheila was familiar with the approach]. Donna agrees to facilitate and asks for more information on the goals of the initiative. Because GoGo has organizational development and facilitation skills in-house and is somewhat familiar with Structural Dynamics, Donna and Sheila decide to engage Art of the Future on a consulting basis. Art of the Future will guide them through the

initiative, coach them on process issues as they arise, and participate in sessions on an as-needed basis.

- **Assembling a Powerful Team**

  January 11, 2010: By the time the executives gather the following Monday, Sheila has distributed the criteria for Team members.[26] The leaders have prepared lists of potential candidates. Sheila calls this selection process a "deep dive" into the organization – looking for the most creative and respected thought leaders wherever they might be positioned. The executives go through an iterative process of proposing, discussing, sorting and resorting prospective Team members to meet the multidimensional diversification criteria. At the end of the meeting, they have compiled a proposed list of 24 internal candidates [and two external], with a second list of alternates.

  ▸ The executives inform their line leaders of the proposed internal Team members. The line leaders approve the choices with a few exceptions. The line leaders then have one-on-one meetings with the candidates. These conversations include a description of the goals of the strategic initiative, the expectations of Team members, how the selection process works, the criteria used, why this specific individual is being asked to join the Team, the time commitment involved and the impact on their jobs and the role they will be playing with their peers, reports and managers. Only three proposed candidates decline to participate. Sheila and Donna decide to reduce the number of internal Team members to 22. They select one candidate from the alternates list who will round out the desired criteria for Team composition.

  ▸ Sheila calls the external candidates. One, Charles Schmidt, works for a major customer. Charlie is interested in participating, as it would expand his experience in strategy and leadership development. However, he feels he is too busy to make the commitment. They discuss his interest in participating through an interview to give his perspectives before the initiative gets under way. Sheila likes this idea and decides to interview other customers as well. The other candidate, Cynthia Loquidara, is a journalist and the author of a popular book on the future of work in the twenty-first century. For her participation, Sheila offers Cynthia publication rights, allowing her to write about the process upon completion of the initiative. Cynthia's schedule does not permit her to participate but she expresses interest in doing a follow-up story. Considering these responses, Sheila decides to limit the Team to internal members and gather as much information as possible from external sources.

  ▸ The confirmed Team members are:

  Jonathan utility vehicle sales representative, St. Louis, 15 years at GoGo, bachelor's in biology, Marquette, 52, white, male

  Sheila Chief Operating Officer, St. Louis, 10 years at GoGo, MBA, University of Michigan, 49, white, female

  Eleanor financial analyst, St. Louis, 15 years at GoGo, CPA, bachelor's in accounting, St. Louis University, 51, white, female

  Juan electric vehicle manufacturing technician, Tijuana, 20 years at GoGo, high school diploma, 45, Hispanic, male

  Kim R&D researcher, St. Louis, 1 year at GoGo, master's in management science, Tokyo University of Science, 40, Asian, male

---

[26] See "Convening the Team" in Part 1 for team-member selection criteria.

GoGo Case

Tom	S.V.P. Human Resources, St. Louis, 5 years at GoGo, master's in organizational development, University of Chicago, 58, white, male
Adriana	logistics manager, Belarus, 3 years at GoGo, bachelor's in fine arts, State University of Moscow, 42, white, female
Alison	administrative assistant, St. Louis, 2 years at GoGo, associate's degree, 33, white, female
Garrett	senior financial manager, St. Louis, 7 years at GoGo, master's in business administration, Washington University, 52, white, male
Carl	fork-lift manufacturing supervisor, St. Louis, 35 years at GoGo, high school diploma, 63, white, male
Darnell	advertising manager, St. Louis, 10 years at GoGo, bachelor's in U.S. history, University of Indiana, 44, black, male
Enrique	IT data analyst, St. Louis, 1 year at GoGo, bachelor's in database administration, University of Maryland, 26, Hispanic, male
Fillipo	product designer, St. Louis, 6 years at GoGo, bachelor's in industrial design, University of Bologna, 43, white, male
Leon	environmental health and safety professional, St. Louis, 7 years at GoGo, master's in English, UCLA, 50, hispanic, male
Heather	transportation logistics assistant manager, St. Louis, 2 years at GoGo, high school degree, 28, white, female
Isaiah	manufacturing line operator, St. Louis, new hire at GoGo, associate's degree, 23, black, male
Jazmin	facilities engineer, St. Louis, 3 years at GoGo, associate's degree, 29, black, female
Katherine	admin coordinator for manufacturing operations, St. Louis, 22 years at GoGo, bachelor's of business management, D'Youville College, 55, white, female
Logan	human resources recruitment manager, St. Louis, 4 years at GoGo, master's in psychology, St. Ambrose College, 35, white, male
Molly	real estate planning manager, St. Louis, 5 years at GoGo, master's in business administration, University of Wisconsin, Madison, 37, white, female
Natalie	Southeast U.S. golf-cart sales manager, Atlanta, 5 years at GoGo, bachelor's in psychology, University of Kentucky, 35, white, female
Quinn	community relations director, St. Louis, 7 years at GoGo, master's in public administration, University of Kansas, 45, white, male
Mia	IT professional, St. Louis, 2 years at GoGo, University of Illinois, master's in computer science, 35, Asian, female
Justin	US golf-cart sales senior manager, St. Louis, 28 years at GoGo, bachelor's in business, St. Louis University, 64, white, male

- **Documenting and Communicating: The Playbook**
Sheila enlists the services of GoGo's web designer, Julie Adams. She asks Julie to be the Playbook coordinator, attending all Team meetings to capture data and put it up in an area of GoGo's internal website dedicated to this initiative. As the initiative progresses, this information will take the form of a Playbook, documenting both the process and the outcomes. Julie and her manager agree to Sheila's request. Julie will be supported by

her assistant for the semester, Rachel, a graduate student in information sciences and learning technology at the University of Missouri. To start, Sheila asks Julie to create the site and post the names, titles and contact information of Team members, a list of selected readings on new forms of work and work environments, GoGo's strategic direction and an outline of the Structural Dynamics process. Julie suggests that they also have a blog so that they will be able to receive comments throughout the process from anyone in the company. Sheila is excited about this idea. Rachel volunteers to moderate the blog. The Team is informed when the site is up and encouraged to review the resources posted there, discuss them with other Team members and share them with their colleagues throughout the company.

- **Modeling a Life-Sustaining Work Environment**
Sheila is mindful of the importance that place and technology play in the success of this initiative. She has asked Jazmin, a Team member from the facilities group, to work with her to identify a space for the Team to meet throughout the process that will allow them to move around easily and to work in small groups comfortably. The meeting space will be available to others when the Team is not using it, so Sheila would also like to have a large wall space in a well-trafficked area to use as a display and work area throughout the project – something like the old "war room" idea used in many creative efforts, but more open and accessible to the organizational input.

They decide to hold the first meeting in a large conference room off the main lobby of GoGo's headquarters. This will be a test of its suitability for use throughout the project. For the display wall, Jazmin suggest a wide corridor on the way to the cafeteria in the same building.

 ## Convening

Session 1 – January 26, 2010: Prescott, Sheila and Donna kick off the process. They are impressed by the range of talents, experience, backgrounds and perspectives represented on the Team. All of these people have demonstrated initiative, an ability to get things done, a degree of creativity and a willingness to engage in open conversation with others. The composition of the Team alone is going to communicate something important to the entire organization.

- **The Importance of Being Life-Sustaining**
Sheila begins by welcoming the Team, introducing herself as the Team leader as well as a member and thanking these recruits for their willingness to tackle the strategic issues that are so vital to GoGo's future. She explains that Team members come from many levels of the organization and that this factor should not interfere with the work they are about to undertake. She provides a summary of the issues and the goals for the initiative. Sheila says that GoGo faces a challenging competitive environment as well as enormous opportunities, saying that:

> "We stand at the threshold of a new era in science, technology and ways of connecting with one another that we haven't yet even dreamed of."

Sheila turns the microphone over to Prescott, who charges the Team with a big responsibility:

GoGo Case

"This is a very exciting and challenging time for our company and for the world. The United States is facing a severe recession. Some of the countries that are emerging as world-class producers and consumers, such as China and India, are ones that we don't know much about culturally, politically and economically. We need to become much more savvy about international business opportunities. We are now playing on the world stage even as we are headquartered here in America's heartland. We've been in business for over 60 years. Looking back, it's been mostly slow, steady growth. Compare that to what we sense lies ahead! Our markets, our products and our processes are changing very rapidly.

"We were not assembled from parts and pieces like one of our golf carts. We started small. We grew by hiring good people and differentiating responsibilities – like an oak emerging from an acorn, forming a trunk, limbs, leaves and acorns for the next generation. This organization is a living system. When we really get that, we will thrive by supporting our people, our organization and our environment. In short, we need to become life-sustaining. When this initiative succeeds, everyone in GoGo will know who we really are and what we stand for...the best products in the service of a better, cleaner, healthier world. We need to build an organization that is operating in concert with the natural environment. We cannot and will not contribute to fouling our water, our land, our atmosphere and our food supply. GoGo can't solve all the world's problems. But I'm very proud of our long history of producing electrically powered vehicles. They are demonstrative proof of our commitment to sustaining the quality of life and the natural environment.

"GoGo needs creative, courageous people at a time like this; people like you. I want you to think about what it will take for GoGo to attract the very best people and keep them – even when they might have other attractive opportunities. Think about what makes you want to stay and contribute to GoGo's success. Whatever it is, let's replicate that!! We need to be a magnet for the best and the brightest talent by providing a work environment that is fulfilling and supportive. By feeling good about what we're doing and how we are doing it, we will attract others who want that kind of experience for themselves.

"None of us has all the answers. But together, we have a lot of them. I'm asking you to help this company figure out how GoGo can best sustain our planet, support our people and build great products that contribute to making the world a better place. As far as I'm concerned, they go together.

"So, we are looking to you, this Team, for a great set of ideas and insights to help us envision the company we'll become in the next decade and beyond. You'll work hard in the next few months, but I think you'll find it to be an exhilarating experience. You'll learn a lot, you'll discover you know more than you thought you did and you'll become leaders in this organization – no matter what your job title is.

GoGo Case

You'll be making discoveries and recommendations that will set us off in the right direction.

"Your thinking will be influential across the organization. I'm standing behind you and your work on this initiative. That said, I'm sure that you're not going to need much more than encouragement from me. What you're going to come up with will be compelling enough that people will get behind it just because of the power of the possibilities you are laying out."

Next, Sheila introduces Donna and explains her role as the facilitator:

"Donna's job is to keep us on track throughout this process. So, she'll be moving us along in the right direction. She's also here to make sure that we have a truly great learning experience. She'll be sure that each one of us feels we can each say whatever is true and important to us. She will continually encourage us to express our best thinking. At the same time, we need to really listen to and consider what is being said. We all want and need to be heard. Donna will be watching to ensure that the Team is keeping to the norms that it will set for itself."

Then Donna provides an overview of the Structural Dynamics process. She explains that it will take place in a series of eight half- or full-day sessions. She says that these gatherings will be scheduled at two-week intervals so that there is time between sessions for connecting with colleagues. These periods between sessions also allow time to absorb the learning that takes place in the Team meetings, apply that learning on the job and prepare for the upcoming session. Team members may engage in additional fact finding, reading or research. Donna says she has found that Team members often come to the following session with fresh knowledge and insights – or sometimes, a realization that a really critical point was overlooked.

Sheila now introduces Julie and Rachel:

"Julie and her assistant, Rachel, are our resources to help us create a Playbook of our process and outcomes that will be available as a communication device to the organization as a whole through GoGo's intranet. We already have a start. In preparation for this meeting, you found the information we posted on our site for this initiative. Julie will be refining and maintaining the site as we go along, and Rachel will be focusing on a blog so that we can get wide participation and input from everyone in the company. This is your communication and documentation vehicle. Be sure that it is complete and accurate. This site will become our Playbook, documenting the results of our efforts and their application throughout the company. I will be looking for a few Team members who can work with Julie and Rachel to help guide the design of the Playbook."

Basic information about each Team member was posted in the Playbook before the meeting. As a "check-in," all members say something about their work at GoGo, their personal interests and their views about any aspect of the natural environment they wanted to mention. They will later have a chance to have these comments videotaped for inclusion in the Playbook. The Team then spends a few minutes appreciating the skills and talents present in the room, e.g., the group of people this project is bringing together and their potential to do something exciting and important.

GoGo Case

- **Articulating the Decision Issue**
  Sheila charges the Team with articulating the fundamental issue at the root of the multiple concerns that have been voiced by the GGEC and others. After a lengthy dialogue, they decide on this formulation of the Decision Issue:

  How should GoGo recruit, house and support the employees we need for continued success?

- **Selecting a Time Horizon**
  Donna asks the Team to consider the amount of time it will take for the decisions made in this process to fully play out. Since their decisions may involve long-term investments and the evolution of social norms, the Team decides that a 10-year period is the right timeframe to consider. Thus, the year 2020 will be used as the time horizon of the scenarios they will develop.

- **Life-Sustaining Environment**
  The Team agrees that the space they are in satisfies their needs. This room will become their home for the duration of the initiative. The Team also likes the idea of being able to use the large wall outside the cafeteria to display graphics and to engage others by encouraging them to contribute to the "Wall." Kim and Fillipo volunteer to organize the materials to be posted on the wall and creating a "look and feel" for the display that conveys a sense of the initiative while inviting comments and participation.

- **Developing Playbook Criteria**
  Julie leads a discussion of the Playbook, soliciting ideas from the Team members regarding how they would like to see it structured, the types of information it should contain and its functionality. Enrique and Adrianna volunteer to work with Julie and Rachel on the structure, content and flow of the Playbook.

Donna leads the Team in a discussion of agreements about "ground rules" or "norms" – how they want to interact during the initiative. Among their ground rules, they include:

✓ We listen to one another without interruption.
✓ We respect everyone's opinion, even when it differs from our own.

Sheila explains the idea of interviewing customers, suppliers and others who care about GoGo. She asks for volunteers to work with her over the next week to compile questions for these interviews. Mia, Logan and Darnell say they would like to participate. The questionnaire will be posted in the Playbook when it is ready, and each Team member should find one external person to interview before the next meeting. They will be able to enter information into the Playbook so that the source remains anonymous and the answers to the questions can be compiled and analyzed before the next Team meeting.

Sheila also encourages all Team members to read widely, to search out experts in a wide range of fields that in any way relate to the initiative and become familiar with their ideas, and to keep their eyes and ears open for information from all kinds of sources.

Charles suggests that the Team needs a name and he is sure that this creative group can come up with something more interesting and fun. Several ideas are bounced around, and the Team decides to call itself "GoGo Forward" – or "GGF" for short.

GoGo Case

Donna proposes that the Team meet on Tuesday mornings from 8:30-12:30 every two weeks. Continental breakfast and lunch will be provided at each meeting for those who can arrive early and stay late. Team members agree with this schedule but want to maintain some flexibility to accommodate unforeseen circumstances. Sheila stresses that this work is a priority for Team members and their leaders, noting that their presence on the Team demonstrates its high priority for the company. She also wants to reserve the right to extend meetings to a full day when it seems necessary as they get into the work. They agree that the next session will be held in two weeks.

Sheila emphasizes that Team members must scrupulously avoid any sense of exclusive membership in a "special" club. For example, they must take care, throughout the process, to explain the meaning of terms if they use any jargon or acronyms pertaining to the initiative.

Homework: The GoGo Forward Team will begin communicating with colleagues in the workplace, including their leaders and direct reports. Once the questions are established, interviews with external experts and stakeholders. Sheila suggests that they start tracking items in the news – reading publications like Time, viewing websites like CNN.com and nytimes.com, or watching news programs, such as BBC News and the PBS News Hour. They will be blogging in the Playbook about any thoughts or information they want to share within the company. They will coordinate with Kim and Fillipo in the design and content to be posted on the "Wall" outside the cafeteria and keep it updated throughout the process. The survey questions for external interviews will be developed and …

GoGo Case

Facts

Exploring Facts

🌐 Role of Leaders: Futurist

During the intervening two weeks, Sheila scans sources of information on how the way people work will be continually affected by economic and social factors as well as advances in science and technology. She also becomes more aware of medical and economic factors that will keep people living and working longer. She familiarizes herself with the demographics that will lead to multiple generations in the workplace. She reads about the latest trends and implications of offshore jobs, including the political repercussions, the impact on jobs at home, how customers are reacting and global wage rate disparities.

She monitors reports of people's attitudes regarding climate change and efforts to care for the natural environment, such as emerging trends in local sourcing, micro-manufacturing, reduced packaging and waste, products made of recycled and recyclable materials, energy efficiencies in construction and building operations, increased use of renewable energy sources such as wind, solar and geothermal and the demand for electric and hybrid vehicles.

Sheila learns that the increase in the number of people working from home is helping to ease traffic congestion and leaving a huge number of offices underutilized. She finds research that supports her own experience: that being able to work anywhere at any time improves productivity but can also lead to overwork, stress and burnout.

She has come across many sources of information on these and related subjects and makes these references available on the Playbook. She knows that Team members are also actively engaged in this exploration of data because they are adding their own sources and suggesting related topics that could be relevant to the inquiry. Her enthusiasm for learning is contagious.

Sheila works with Team members to craft the questions to be asked of external players, and she monitors the answers as they are posted in the Playbook.

🌐 Scanning Events

Session 2 – February 9, 2010: The Team, now known as GoGo Forward, shows up eager to get to work. Today they will identify forces that may impact the Decision Issue that GGF identified in the Convening session.

Donna reviews several criteria for events with GGF. Events must:
- ✓ Relate to the Decision Issue
- ✓ Occur within the time horizon [in this case, 10 years]

GoGo Case

- ✓ Be variable, i.e., change over time such as increase, slow, improve, etc.
- ✓ Be outside the control of the organization

Contradictory events are fine; they can be woven into alternative scenarios. The Team considers the following framing question to focus the ensuing exploration for relevant events:

What events, factors, behaviors, conditions or outcomes would – if they occurred in the next 10 years – have a significant impact on how GoGo recruits, houses and supports the employees we need for continued success?

To encourage GGF to think broadly and look for events from a range of domains, Donna asks members to think of one or more in each of the STEEEPA categories. She provides each GGF member with a felt-tip marker and a worksheet like the one shown here that has the domains listed with three color-coded 3" by 5" sticky notes next to each one.

Donna instructs GGF to write each event concisely and legibly, in large letters, one per sticky note. As GGF members spend a few minutes thinking individually and jotting down ideas, she sets up seven flip charts around the room. As a heading on each, she writes one of the STEEEPA domains.

After GGF members have generated events individually, they are asked to form equal-sized domain groups and gather at one of the flip charts.

One by one, the domain group members post sticky notes related to their domain, clearly identifying its relevance to the issue. The group members then discuss the events that have been posted. They do not discuss the likelihood of an event. They review events only by the four criteria: relevance to the Decision Issue, capable of occurring within the timeframe, variable and outside GoGo's control. They include any event meeting these criteria. Additional events that come up in discussion can also be included.

After 10 to 15 minutes, the groups rotate clockwise and review the events in the domain on the next flip chart. If they have jotted down events on their sticky notes that are not yet posted, they add them and any others that occur to them.

After seven rotations [about 90 minutes], the groups are back at the domain where they started. The Team has generated a large number and a wide range of events. The domain groups review what has been added to their flip chart, discard any duplicates and cluster similar ideas.

GoGo Case

Each domain group reads several of their events aloud:

Societal
- Many C-suite executives are imprisoned for misdeeds of their corporations,.
- Fathers gain equal parental rights including child support from mothers.
- More employees are caring for elderly parents at home.
- Many employees choose not to retire.
- Employees opt to live close to their workplace, commuting by foot or bicycle for health, cost and environmental reasons.

Technological
- Hydrogen fuel cells power vehicles.
- Video telephony is ubiquitous.
- WiFi is freely available in most U.S. and Asian cities.
- Mental telepathy is demonstrated as a means of communication.
- Electric jitneys are popular as low-cost, convenient urban transportation.
- Organizations introduce full-body screening for all employees as a security measure.

Economic
- Business travel costs skyrocket.
- The economies of several countries and states collapse.
- Housing units shrink in size.
- People use more communal services to save money.
- Wage rates equalize globally.
- Compensation is based solely on experience, education and skill.

Educational
- The number of on-line degrees exceeds campus-based programs.
- Libraries partner with top colleges to offer degrees using both on-line and in-person education.
- The free public library becomes an artifact of the past as ongoing municipal budget crises cause fees to be imposed on all library usage.
- Design courses are required in Europe for business students.

Environmental
- Carbon taxes limit the number of miles people are willing to drive.
- Communities are self-sufficient in growing food, making goods and disposing of waste.
- Rising sea levels shift many population centers.

Political
- Government regulation and populism increasingly polarize.
- Terror incidents plague many places once considered safe.
- Global governance organizations weaken.
- U.S. mandates that public transportation meets European standards.

Aesthetic
- Studies prove that natural light in the workplace improves productivity.
- Organizations compete for talent based on the appeal of their environments.

- Abandoned railways are redesigned for walking and cycling commuters.
- Well-designed products command a premium price.

- **Safe Bets: Certainties**

    GGF members review the work on all the flip charts and then come back together as a whole group. Donna asks them to consider whether any of the events they have been discussing are absolutely certain to occur – events that will happen, that are not contingent on any other factor or event, that are inevitable.

    Garrett proposes, "Skilled labor in the U.S. will be scarce." He knows that one-third of Americans will be over 50 years old by 2012, even though the impact of this development on the workforce hasn't received much attention in the current recessionary environment. To reinforce his point, he shows the Team an article from *The Economist* on how to survive the "silver tsunami" [Schumpeter, 2010].

    Others point out that a shortage of workers depends on other forces in play, such as technology innovations, immigration rates and the number of jobs created or eliminated. The group agrees that there might be shortages in some fields, but not on whether there would be any general problem with the supply of talent in the workforce. They accept as a certainty that, "The median age of the U.S. and Belarus workforce will continue to increase over the next 10 years."

    After more discussion, GGF decides there are no other events that they feel confident calling "certainties." Donna moves their one "certainty" to the side and will remind them to include it in all the scenarios of the future they will be developing, because they believe it is "certain" to occur.

- **Events Deserving Close Attention: Critical Uncertainties**

    GGF discusses which of the events are most uncertain, i.e., the most volatile. Uncertainty refers to the degree to which it is unclear if such an event may or may not occur. Donna then turns discussion to which of the events are most critical to the Decision Issue. Criticality refers to the degree of impact the event would have were it to occur. This discussion is not intended to lead to any outcomes; it is simply to consider the events with the lenses of uncertainty and criticality.

    Each Team member receives five red and five yellow "dots." Team members "vote" by placing their dots next to events.

    - 🟡 Yellow dots for the most uncertain events
    - 🔴 Red dots for the most critical events

    After all the dots have been placed, Sheila leads a discussion to review the results of the voting and discusses those events with the highest number of dots.

    A couple of events have clearly received the highest number of "votes":
    ➡ Computer science and engineering graduates are in high demand In the U.S.
    ➡ Environmental concerns are causing many countries to pass strict regulations.

    Several other events have received many dots, with a high preponderance of either red or yellow dots. Adrianna asks, "Is there any reason to favor red or yellow dots?" Donna offers the following thoughts:

GoGo Case

- Red dots could indicate that the event is <u>highly critical</u> but not <u>highly uncertain</u>: it may be within the control of the organization.
- Yellow dots might designate a <u>highly uncertain event</u> but one that is not <u>highly critical</u>: its relevance to the Decision Issue may be low.

To select the critical uncertainties, GGF discusses the degree of control GoGo exerts over the events with a large number of red dots and the degree of relevance to the Decision Issue of the events with a large number of yellow dots. The discussion results in the selection of three additional events:

➡ Most workers commute to corporate offices no more than once a week.
➡ Economic recession/depression continues through 2020.
➡ Women's rights are trampled in many countries.

- **What's Changing: Variables**
  GGF determines the underlying variable for each of the events they have selected:
  ▸ supply of Skilled Workforce
  ▸ amount of Environmental Regulation
  ▸ percent of Remote Workers
  ▸ degree of Global Economic Stability
  ▸ value placed on Human Rights internationally

  These variables are the critical uncertainties that the team will be working with as they proceed.

To wrap up the first session, Donna asks them to think about the events they identified, the critical uncertainties they selected and the variables they will be working with. She asks them to consider the questions:

➡ Did we cover a broad range of events that could affect the way GoGo recruits, houses and supports its workforce?
➡ Did we leave any critical events out?

Homework:
Watch for these and other variables "in action" in the news.
Communicate...
- Blog about this experience.
- Share information with your colleagues both formally and informally.
- Get on the agenda at staff meetings.
- Create other opportunities to address a group.
- Stay in touch with your teammates.
- Communicate directly with your leaders.
- Feed back observations and thoughts in the Playbook.
- Update the Wall.

Julie posts the events, critical uncertainties and variables in the Playbook along with a description of the scanning exercise.

GoGo Case

🌀 Recognizing Patterns and Structure

Session 3 – <u>February 23, 2010</u>: When GGF reconvenes, Donna asks the members to comment on their thoughts and experiences in the intervening weeks. The participants report a great deal of excitement within GoGo for the work they are doing. Several events have been proposed through the Playbook that were not discussed in the last session. GGF spends some time considering them. They decide that, while relevant, these events are not among the most critical or the most uncertain. They will be added to the other events.

Donna has placed each of the critical uncertainties [expressed as variables] in the center of a flip chart. Team members form small groups to work on each of the five. The task is to identify variables that impact the critical uncertainty and those affected by the critical uncertainty [i.e., the cause and effect relationships]. Here is the pattern that the group working on "Supply of Skilled Workforce" came up with:

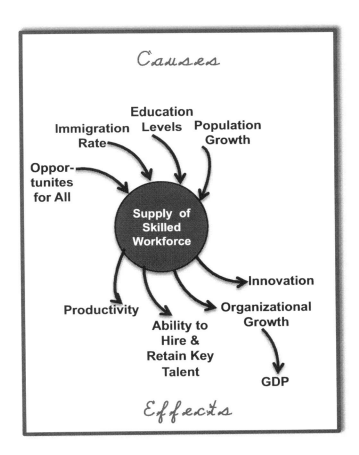

The purpose of this step is to identify a few of the key variables useful to the analysis. Other variables will emerge as the initiative continues.

141

GoGo Case

🔵 The Heart of Structural Dynamics: The Scenario Game Board

GGF is ready to start thinking about organizing the information they have amassed in a way that will lead to understanding. They continue to work in their small groups organized around the variables.

- **Pace of Change: Abrupt or Gradual**

  Donna asks each group to consider the forces that impact its critical uncertainty and discuss how change could occur gradually; and how it could occur abruptly. For example, the supply of skilled workforce would change abruptly if a dramatic opening of immigration limits for particular skills occurred. Other variables would more likely generate gradual change.

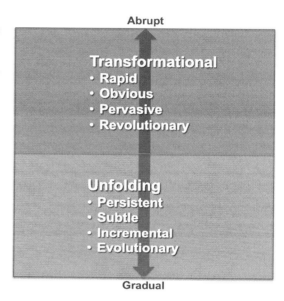

- **Response to Change: Reactive or Creative**

The groups then consider the possible responses to these changes; why some people might respond with fear and others with hope. In our example, new entrants into the workforce could evoke fear in long-term employees if they seen as a threat to the jobs of existing employees. Others might experience those entering the workforce as providing the opportunities to gain new knowledge.

- **Integrating the Dimensions**
  Donna explains how these two dimensions, when used together, create four quadrants. Each quadrant provides the foundation for a unique scenario of future possibility.

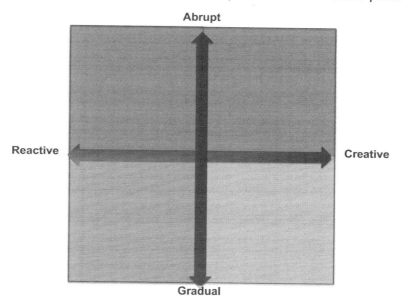

- **Scenario Archetypes**
  Donna introduces the archetypal scenarios to GGF. She explains how the cause and effect variables they have identified could interact to create different futures.

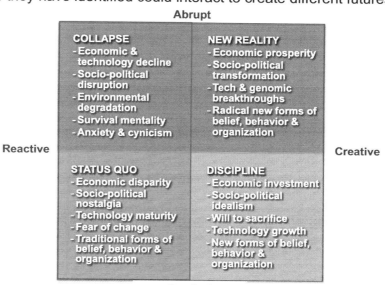

GoGo Case

- **Scenario Game Board** ©2010
  To set up the detailed conversations about each of the critical uncertainties that will follow, Donna continues to use "Supply of Skilled Workforce" as an example. She places it in the center of the diagram and asks GGF to give examples of how this critical uncertainty could generate very different future conditions.

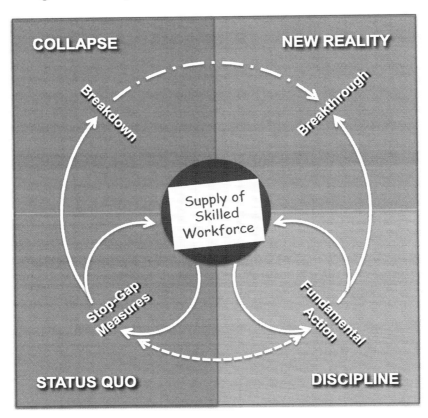

The Game Board

Enrique says he can imagine a situation in which there wouldn't be enough information technology specialists in some parts of the world to handle the complicated logistics or sophisticated equipment that so many industries depend on. Many systems like telecommunications could just stop working, which would affect many other things including, at some point, social, economic and political stability.

Kim, who has been closely following demographics in China and Korea, thinks that it's easy to imagine a world where there are plenty of engineers and technicians from the Asian countries to handle complex assignments around the world, as long as nations have immigration policies that encourage a distribution of these skills.

Adrianna thinks that many companies and countries will continue with what they're doing simply because they don't know what will be needed in the future and can't train people for something they don't know anything about. "It's certainly the way most people operate."

Logan says that things are changing much more rapidly than any of us realize and that all sorts of innovations are much closer to becoming reality than we may think: "If the need is great enough, we'll discover that there is an enormous amount of talent ready to break through old barriers and stereotypes and get to work."

"Great stuff!" says Donna. "That's the sort of imaginative thinking to bring to the next task." At this point, GGF goes into their small groups around the critical uncertainty each has been working with. We will continue to follow the group working with the critical uncertainty, "Supply of Skilled Workforce." They now consider the interaction of the causal relationships that could lead to each of the archetypal scenarios.

▸ **Maintaining the Status Quo**

If the supply of key skilled workers were to shrink, the group determined that organizations could react by increasing the number of hours employees are required to work, raising pay to attract new workers or seeking to influence the supply of temporary immigrant workers. These actions would have the effect of keeping things in balance, so that supply would meet demand.

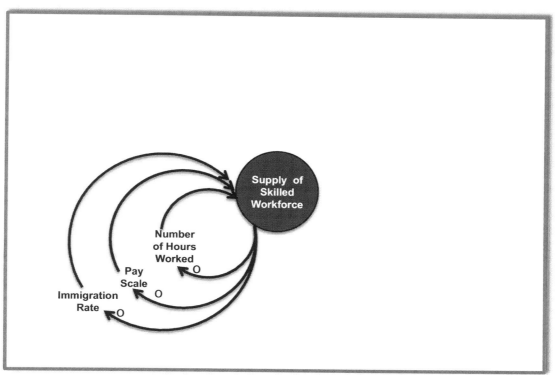

STATUS QUO

GoGo Case

▶ **Exercising Discipline**

Alternatively, organizations could invest in administrative, managerial and communication systems to increase their access to remote workers [locally, nationally and globally] to broaden the area in which they seek people with the talents and skills they need. They could also hire those who have traditionally been excluded [women, handicapped, minorities, self-taught, etc.] in these professions. In addition, organizations could support education and training to grow talent. These actions would serve to increase the supply of skilled workers but may take longer to implement.

The group adds the dotted arrow to indicate that both types of responses might be used at the same time. For example, a temporary increase in immigration rates may be necessary until local talent is educated or until systems are in place to support remote work. An oscillation takes place between short-term approaches to keep things going while new approaches are developed and implemented.

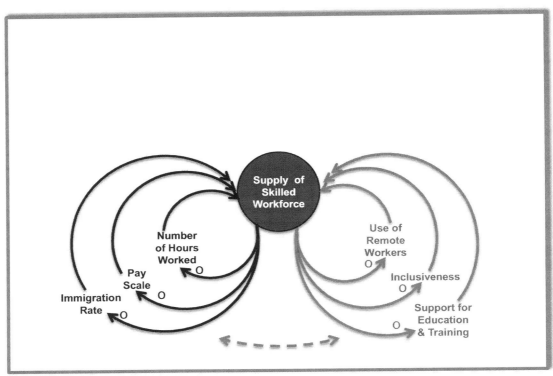

STATUS QUO   **DISCIPLINE**

GoGo Case

▸ **Stepping into a New Reality**
The use of remote workers strengthens an organization's ability to continue operations in the face of a business disruption [i.e.,s increases business continuity]. Inclusiveness and the use of remote workers also increases diversity in the workforce. Support for education and training leads to higher education levels. Increased diversity combined with higher education levels leads to more innovation and enables a true meritocracy to develop. Business continuity and innovation improve productivity. A meritocracy with high levels of productivity leads to organizational success. This combination of factors will yield an energetic, new reality significantly different from the present state.

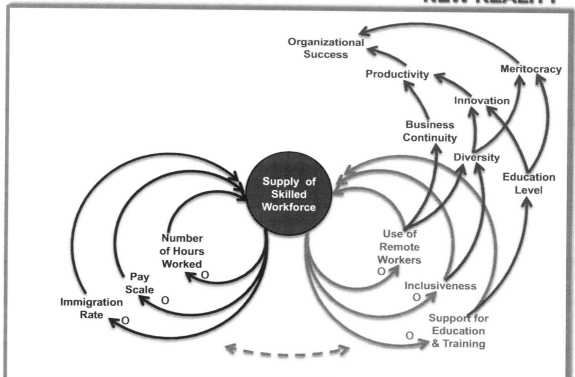

GoGo Case

▸ **Experiencing Collapse**
The group notes that holding on to the Status Quo may lead events to tip too far in an unanticipated direction.
- A temporary increase in the immigration rate adopted to maintain stability may be hard to reverse, leading to misunderstanding and conflict within the workforce.
- A general rise in pay may spark an inflationary wage war.
- Increasing the number of hours employees are expected to work can lead to burnout and a sudden, dramatic and irreversible loss of productivity.

The group puts the dotted line to indicate that conflict, inflation and burnout would make it extremely difficult, if not impossible, to achieve organizational success.

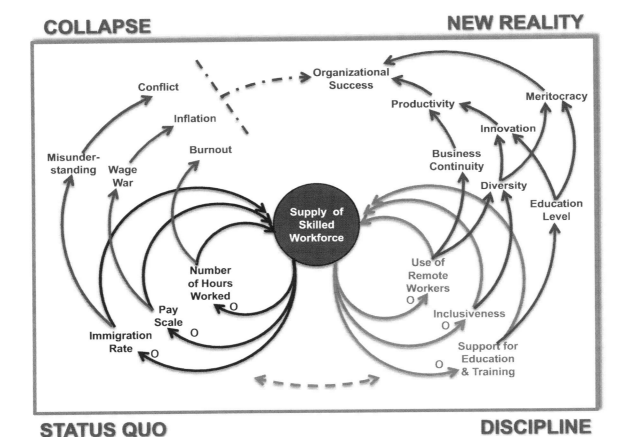

GoGo Case

🌐 Exploring Deeper: Investigating Mental Models, Myths and Metaphors

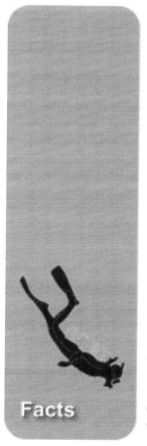

Facts

The GGF groups know that that there are two kinds of forces in the structures they are creating: physical and human choice. They now focus their attention on the human choices. Donna challenges them to surface their assumptions. She asks each person to reflect on any cultural assumptions they think might be built into the structural diagrams they have created.

After a moment of individual reflection, GGF members discuss their thoughts with the person next to them. In 10 minutes, the pairs join their groups and continue the discussion. After another 10 minutes, Donna asks the groups for insights they would like to share. When each group has had a chance to speak, she asks if anyone feels they have another point they would like the Team to discuss. Several members add their perceptions to what has already been said. GGF starts to see that what is obvious to one person may not be obvious to another. This is one of the many points in the Structural Dynamics process where the diversity of the Team is a powerful asset to accomplishing its objectives.

One point that emerges is that quick-fix solutions can be more attractive than approaches that take more time and investment of resources. The reasons for this may not be entirely logical, as they may reflect some powerful underlying mental models. For example, many traditional command-and-control leaders believe that the workforce needs to be watched to be productive. Anyone holding this point of view would steer away from the use of remote workers, perhaps without even being aware of the reason. Tom makes the observation that cultural myths and metaphors can keep the status quo in place long after it is appropriate to current reality.

Homework: To test its assumption that remote work leads to increased productivity, GGF decides to research the experience of other organizations that use remote workers. They will blog about their findings in the Playbook and present them in the Executive Briefing. GGF continues communicating, blogging and updating the Wall. They will review these diagrams with their colleagues.

GoGo Case

🧀 **Executive Briefing: Facts**

March 9, 2010: The Executive Committee gathers for a two-hour session to hear the results of the fact-gathering phase of the work. Those present are the GGEC, GGF, Julie, Rachel and Donna. Sheila opens the meeting by reminding Prescott that he wanted the GGEC to be kept informed of progress and insights along the way. GoGo Forward Team members also feel it is necessary and important to check in periodically. Sheila points out that GGF members are not only communicating among themselves and to this group of decision makers, they are also keeping their colleagues informed of the process on a regular basis between GGF sessions and soliciting their thoughts and comments both through individual contacts and through the PlayBook.

Prescott makes an opening comment:
> "This is exciting! We are looking forward to hearing your thinking. We are not here to judge. We're all in this together. You are doing valuable work for the company. So, if you've been worrying about the so-called Big Cheeses, please don't. We're here to learn how to move GoGo into a prosperous future."

Garrett describes how the GoGo Forward Team has defined the Decision Issue as:
> *How should GoGo house and support the employees we need for continued success?*

GGF members felt that this framing would lead them to discover a great deal about becoming a life-sustaining organization.

Leon then recounts how GGF arrived at a set of variables that are critical to this issue and are also highly uncertain. He presents these and leads the discussion of each.

Eleanor describes how GGF identified forces that affect each of these critical uncertainties and then what each of the critical uncertainties impacts in turn. This process allowed them to see patterns start to emerge – things that happen over and over. She shows the diagrams that GGF developed for each of the critical uncertainties.

Juan explains how each of these critical uncertainties could play out in different ways using the archetypal scenario framework as a guide. A member of the group that developed each structural diagram walks the Executive Committee through the thought process used to arrive at these diagrams.

Tom summarizes GGF's understandings:
- There are a number of macro, big-picture forces in play that have an impact on how Go-Go should house and support the employees needed for continued success [the Decision Issue].
- These forces act in dynamic relationship with one another.
- As a result, there are many ways in which the future could emerge.
- Knowing how these forces interact provides useful insight into where things may be heading.
- Being aware of the dynamics in play, GoGo can become less dependent on past experience and more open to considering a wide range of possibilities going forward.

- By scanning for relevant variables, monitoring emerging trends and knowing the significance of these events, GoGo has a leg-up on its competitors who may be blind-sided by changes they had not anticipated.

The reaction from the Executive Committee:
- "Some of the critical uncertainties GGF has identified have been obvious to us but others we hadn't considered. They could have a tremendous effect what we do and, ultimately, on our results. It's important to be aware of them and monitor what's going on."
- "When we look at the second- and third-hand effects you've identified in the structural diagrams, they are not obvious or intuitive. It helps to have the diagrams to see the implications of changes in our business environment, the consequences of our actions and those of other players."

Denise expresses a concern by asking, "You will be sharing your experience with people who have not been directly involved in this process. How do you respectfully communicate your insights and learning without creating a sense of exclusivity?"

Molly, says she has been thinking about this issue and describes her communication plans:
- "I'm going to send an email to everyone in Real Estate and Facilities Management describing the Decision Issue and why we think it's so important.
- "I'll call several people I know in the company who are influential in their areas and describe our work to date and ask for reactions. They are line leaders like me, and they really know how to get things done. I'll ask them to help me set up meetings to communicate and get responses, to make sure we're on the right track. I plan to keep the conversation focused on the structural diagrams using questions like, 'Have I explained it clearly? Did we leave out any relevant factors?'
- "I've already made a couple of blog entries and that was a great thing to do because it helped me organize my own thinking."

Prescott and other executives are pleased about these kinds of plans. They are highly supportive of GGF and their work to date. They promise to make it clear to everyone in the organization that they are solidly behind this initiative.

GoGo Case

Discovering Options

🌑 Role of Leaders: Strategist

As GoGo's chief operating officer, Sheila is responsible for developing forward thinking throughout the organization. She wants GoGo to be way out in front of industry trends. She is known throughout the company as a "quarterback" who is not afraid to "call the plays." She realizes that the Structural Dynamics process gives her a powerful tool to enable her to be more effective by involving the whole organization in GoGo's future. Sheila hopes that the process will generate powerful strategic insights. She is excited to be developing concrete steps to bring great people into GoGo as full-time employees, part-timers or independent contractors.

In preparation for the next GGF session, Sheila and Donna divide the names of the members into four scenario groups. They do so carefully to assure a mix of knowledge, experience and perspectives while maintaining a diversity of disciplines and levels in each scenario group. They realize that any GGF member who strongly doubts the possibility of a particular scenario could benefit from being part of that scenario group. Strong supporters of a scenario should be given the opportunity to "try on" another, perhaps diametrically opposed, scenario. This is a way to build the company's strategic thinking "muscles."

Sheila and Donna are aware, for example, that Alison strongly believes that global dependence on fossil fuels will continue and inevitably lead to environmental disaster. They decide to put her in the New Reality group so that she can explore what that situation would be like. They also know that Katherine, who has been with the company since she graduated from college, is a strong believer that GoGo's long history of success demonstrates that the company knows what it's doing and should just keep doing it. Sheila and Donna decide to put Katherine in the Collapse scenario group.

Sheila suggests they reserve breakout rooms for the scenario groups. Donna knows that the buzz created by working in a "beehive" of activity in the same room will energize the process. They decide that the scenario groups will work in the corners of the large room they have been meeting in.

GoGo Case

🌐 Articulating Scenarios

Session 4 – March 23, 2010: GGF members are juiced by the progress they are making and the reception they got at the Executive Briefing. They feel that the executives really "got it." They also received some good comments from colleagues in response to their conversations and their blog entries. As a result, GGF members have a number of suggestions for modifications to the structural diagrams. They are eager to get on with the work.

Sheila congratulates GGF on their success so far in uncovering dynamics that affect GoGo's future. Her opening remarks reflect her enthusiasm for what has been accomplished:

> 'We'll be spending time today living in alternative futures and seeing what they tell us about GoGo's prospects. Doing that demonstrates a willingness to experiment with possibilities. In this session, we'll be honing our skills as strategists. You may find yourself in a scenario world today that makes you feel uncomfortable. Think of it like going to see a good movie. You may have to suspend your disbelief at times to let in the possibility of things occurring in ways you might not expect."

GGF members gather around the set of structural diagrams they built regarding the critical uncertainties. They present potential changes that have been suggested by GoGo employees in the past few weeks. They discuss each suggestion at some length and make those changes and additions they agree upon to the structural diagrams. They remember, though, to include only those variables that add more clarity than complexity.

Donna then describes the next steps:

➡ Link the dynamics between the diagrams to create a Structural Dynamics Model
➡ Use the Model to describe highly distinct scenarios of future possibilities.
➡ Craft scenario narratives.

Donna leads a discussion of how these diagrams interact: how one critical uncertainty affects the others. GGF members look for common variables that appear in two or more of the diagrams. They sketch links between these variables across the diagrams and discover how a change in a common variable affects more than one critical uncertainty. For example, the variable "Inflation" shows up in both the "Supply of Skilled Workers" and "Global Economic Stability" diagrams.[27] They capture these relationships and discover the implications for both critical uncertainties. GGF makes the interconnections between diagrams explicit. Fillipo says, "This is like playing 3D Chinese checkers. I had a lot of fun with that as a kid. I was the champ."

These diagrams and their interconnections become the Structural Dynamics model of the system surrounding the Decision Issue. After being tested from a variety of perspectives, the model becomes a rigorous picture of the system of interrelated forces that are shaping GoGo's ability to recruit, house and support the workforce it needs. GGF looks at it and understands the big picture surrounding the decision issue.

---

[27] Inflationary or deflationary pressures drive dynamics in both diagrams, creating a causal connection between them.

GoGo Case

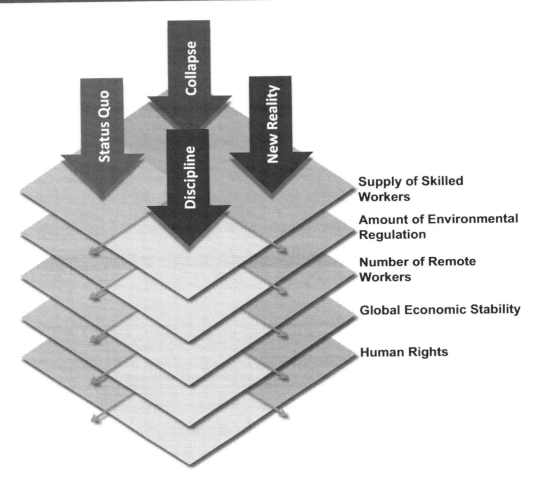

The Structural Dynamics Model

While GGF members feel good about what they have accomplished, they know that the work of constructing the model is never really complete; it is part of the ongoing learning process. There will always be other factors in play that they will want to add to the model as their work continues. If they realize that other factors need to be included because of their relevance to the Decision Issue, they will add them.

After a break, during which copies of the model are made, the groups work on each of the scenarios. The names of the members of the four scenario groups are listed on flip charts in the corners of the room. Copies of the model [all of the diagrams related to the critical uncertainties and their interconnections] have been provided in each corner work area. GGF members find their names on the flip charts and settle in with the other members of their "world." They will work in these groups in this session and the next. Donna asks them to place themselves in the year 2020 [10 years out, based on the selected time horizon]. She tells each of the scenario groups to assume that their scenario world has happened.

Each group explores its world by referring to the dynamics associated with its scenario in the model [i.e., the New Reality group will focus on the connections in and leading to the upper right quadrants of the diagrams; the Status Quo on the lower left, etc.]. They spend time becoming familiar with the causal relationships depicted in the model and discuss their implications. For example, the Discipline scenario group finds that, in their world, innovative approaches are being taken to seemingly intractable issues. The group also notes that the sacrifices required to make changes can engender anger in some and a tendency to pull back toward Status Quo.

- **Living in the Future**

    To get a real sense of life in their scenario, each group visualizes itself in the year 2020 and considers the following:

    <u>What is it like to live in this world?</u>
    Each group details the ways in which its world affects people's lives, the nature of work, child-rearing practices, the quality and type of public institutions, transportation, etc. They consider how the world affects their personal choices and the advice they give their kids. They describe personal life in this world in full color.

    <u>What is it like to work in this world?</u>
    The groups consider where they work, how they communicate, how their performance is measured and how they are compensated. They discuss how much they find their work meaningful and fulfilling.

    <u>How did this world come into being?</u>
    To determine how their world evolved from where it was in 2010 to where it is now in 2020, each group develops *a chronology of events* starting in 2020 and moving backward to 2010. The Collapse group finds that a combination of pandemics and conventional warfare got them to this point.

    Donna has brought along video cameras, props and materials for the groups to work with. She encourages them to be innovative and have fun with the next two activities. She asks the groups to create:

    ▸ A video or visual display illustrating three to five stories of life in the year 2020 in their scenario world
    ▸ An artifact [flag, news headlines, civic motto, description or visual portrait of a working person, etiquette manual, social hero or heroine, etc.] that typifies their scenario at the mythic level

- **GoGo in the Scenario**

    Donna now asks the scenario groups to move beyond living in the scenario to being members of the GoGo organization in the scenario. She has prepared roles for the group members to play to ensure that a variety of perspectives are represented [i.e., HR, IT, Finance, etc.]. When everyone has chosen a role, she asks the groups to do the following:

    - Imagine the key features of the work environment in terms of personal interactions, the setting, reporting systems and management methods.
    - Develop a hiring and appraisal process, e.g., the competencies and personal characteristics most valued in employees.
    - Identify the factors affecting the ability of the organization to hire and retain the talent it needs, e.g., the role the organization's products and services play.

GoGo Case

- Consider what is special about the organization that attracts talented individuals and makes them want to stay.
- Describe how the organization relates to its socioeconomic environment in the scenario, with particular attention to how it competes for creative talent.

This is not the time to develop strategies. That comes next. Here the groups are simply describing working life at GoGo in 2020 in their scenario.[28]

- **Naming the Scenarios**
  The groups give their scenarios short, catchy names that trigger mental images of their worlds. After some deliberation, the scenario groups come up with the following names:
  ★ We can work it out! [Status Quo]
  ★ Reboot [Discipline]
  ★ Quiet Power [New Reality]
  ★ Tough Times [Collapse]

- **Scenario Narratives**
  At the end of the session, the scenario groups review what they have learned about their worlds and capture the essence in bullet form. Here is a sample for each scenario:

  In Status Quo
  - Societal:
    - women continue to gain increasing levels of responsibility in greater number
    - periodic, face-to-face meetings required to maintain trust and teamwork
  - Technical:
    - modest investments to maintain IT infrastructure and product development
  - Economic:
    - sales for most product lines are flat; growth in golf cars in China
  - Educational:
    - college grads face a shortage of professional positions and take interim jobs to get healthcare and pay the bills
  - Environmental:
    - conservation concerns focus on cost savings
  - Political:
    - India, China and other expanding markets require organizations to manufacture in their countries with a local workforce to market their products
    - resource and ethnic tensions create difficulties in recruiting non-nationals in every region that GoGo operates
  - Aesthetic:
    - investment in museums, sports facilities and entertainment venues is strong
    - customers appreciate good design and are willing to pay a premium for it

---

[28] One GGF member said that when she participated in these exercises, it brought her face-to-face with her dislike of working at home, apart from her colleagues. It was an experience that made her examine her values and get in touch with what was important to her in her life and her work.

In Discipline

- ☐ Societal:
  - ▸ strong controlling societies support creative approaches to improve economic, social and environmental conditions
- ☐ Technical:
  - ▸ strategic alliances to support the development of renewable energy economy
  - ▸ significant investments in IT infrastructure
  - ▸ widespread support for virtual reality, holographic imaging and other means to increase communication and interaction among dispersed workforce
- ☐ Economic:
  - ▸ slow, sustained growth funds heavy investments in national, academic and corporate research and development efforts
- ☐ Educational:
  - ▸ funds fuel-cell research projects at several universities
- ☐ Environmental:
  - ▸ water and air are cleaner in most parts of the globe
  - ▸ some endangered species are increasing in number
- ☐ Political:
  - ▸ strict regulation and carbon caps at state, national and international levels have slowed emissions and pollution
- ☐ Aesthetic:
  - ▸ careful attention to functionality, human factors [e.g., ergonomics], visual appeal and use of materials has earned many new product design awards which boost sales

In New Reality

- ☐ Societal:
  - ▸ age of workers has extended to both the very young (learning while earning) and the very old [many well over 100]
  - ▸ they are mostly free agents – highly distributed, independent and creative
  - ▸ many have fierce loyalties to organizations aligned with their ethics and values
  - ▸ travel and tourism are popular
- ☐ Technical:
  - ▸ solid-state hydrogen powder powers vehicles and machinery
  - ▸ speech recognition technology provides simultaneous translation for communication around the world
- ☐ Economic:
  - ▸ manufacturing and food production is done locally, strengthening regional differences
- ☐ Educational:
  - ▸ lifelong learning; value placed on knowledge and capability rather than degrees, status and achievement
- ☐ Environmental:
  - ▸ the environment is in recovery; strong values prevent further abuses

GoGo Case

- Political:
  - the company merged the initial foothold it had in Africa through golf with its approach to diversity to establish a network of good relations across the Sub-Sahara
- Aesthetic:
  - workplaces, homes and public areas have developed in harmony with nature
  - non-aesthetic products and places rarely emerge
  - people have developed a fluidity of motion, speech, dress and interaction

<u>In Collapse</u>

- Societal:
  - traditional gender roles have disappeared in the workplace
  - workers are willing to respond to any opportunities that show up
  - GoGo employs those people who work well together to support its commitment to being life-sustaining
- Technical:
  - little investment or innovation
- Economic:
  - GoGo is a much smaller company: the golf line has disappeared; the utility and fork-lift businesses are smaller; service has become a major product line
  - GoGo has sold most of its real estate assets and leases space in its headquarters facility
- Educational:
  - high schools focus heavily on trades and channel academically oriented students into professions like law and medicine; many fewer institutions of higher learning
- Environmental:
  - less carbon consumption and less pollution
  - recycling has become more like salvaging, as things that were once plentiful are now scarce
- Political:
  - after a number of inconclusive wars, nations have less power; international corporations are able to set their own terms and conditions in most countries
- Aesthetic:
  - good design gives products an edge over competition

<u>Homework</u>: Using these bullet points, the scenario groups will collaborate during the upcoming week in writing two-page narratives of their scenarios that capture their essence. Donna requests that the narratives be posted in the Playbook at the end of one week so that GGF and the rest of the organization can read the complete set before the next session. Once the narratives are complete, GoGo's Media Services will collaborate with the scenario groups to produce seven-minute video clips of each of the scenario worlds. These will be used for multiple purposes, including presentation to the GGEC in the next Executive Briefing, shown to a worldwide meeting of GoGo employees and posted in the Playbook.

Identifying Options

Session 5 – April 6, 2010: Between sessions, besides writing the narrative, the GGF members have all been doing a great deal of thinking about working life at GoGo in their scenarios. They have engaged their colleagues in generating a sense of the implications for GoGo and its employees. They have deepened their understanding of the joys and the restraints of life in their settings. They have prepared materials to embellish their "worlds." One scenario group has made a video clip containing news stories of the day; another has examples of entertainment from their world; others have artwork; one group arrives in proper attire for their world. Donna reviews the agenda for the day:
- Determine strategies that work in each of the scenarios.
- Test these strategies across the scenarios.
- Identify robust and contingent strategies.

- **Effective Action in Each Scenario**
  Given what they know about the world and day-to-day life at GoGo in 2020, each scenario group considers these questions:
  → What are the critical skills we need to have in our workforce at this time?
  → How will we find them, i.e., where are they likely to be?
  → How will we attract and support them?
   ‣ What type of physical work environment, if any, will our creative talent require?
   ‣ What percentage will come to one of GoGo's facilities versus those working at home or on the run?
   ‣ Are they globally or locally dispersed or co-located? Or some combination?
   ‣ What kind of technology do they need?
   ‣ What are their transportation and communication options?
  → What sort of considerations do we need to give to employees' personal needs, i.e., child care, elder care, physical requirements, etc.?
  → How do we manage expenses while simultaneously attracting top talent?
  → How do we mange productivity?
  → How do we create a sense of organizational identity and community?
  → What are competitors doing? How do we stack up? Is what they are doing right for us?

  Based on the answers to these questions, each scenario group develops of set of strategies [usually two to four per scenario] that will help GoGo get to where it wants to be in 2020 in the scenario group's particular world. As GGF goes on a break, the scenario groups give their strategies to Donna so that she can prepare the strategy matrix for use later in the session.

- **Stress-Testing Strategies**
  When the scenario groups reconvene, it is time to see how the strategies developed for each scenario play across all the scenarios. Each scenario group picks one or two "ambassadors" to visit each of the other worlds in turn. They will go as learning agents to each of the other scenario worlds to exchange information.

G₀G₀ Case

The other members of the scenario groups remain in their own worlds and prepare to welcome the ambassadors from the other groups. Their job is to explain their scenario to the visitors.

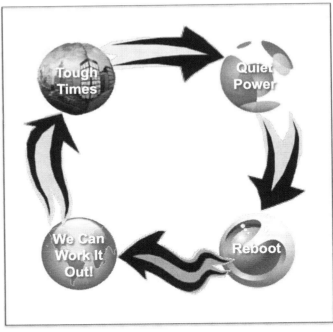

When the first ambassadors arrive, the hosts explain what their world is like and show them artifacts[29] of their world. They tell the ambassadors how events unfolded to create the conditions that now exist. They also describe life and work at GoGo in 2020 in this world.

The ambassadors explain the strategies developed in the world they come from. It is important that the host scenario group does not discuss their own strategies! They listen to the ambassadors' strategies and discuss how those strategies would work or not work in the world the ambassadors are visiting. The ambassadors record the evaluation of each of their strategies on a worksheet. [See sample below.]

Strategies \ Scenarios	We can work it out!	Tough Times	Reboot	Quiet Power
We can work it out! Strategy 1	+			
We can work it out! Strategy 2	+			
We can work it out! Strategy 3	+			

Ambassadors' Worksheet

---

[29] A Status Quo group member spotted this bumper sticker and thought it captured the tone of their world. The group used it as an artifact to convey a sense of their world to visiting ambassadors.

The ambassadors then rotate to the next scenario world and repeat the process. This continues until the ambassadors are back in their own world. The ambassadors report their findings to their own scenario group. The group considers the results of the stress tests and discusses the implications of this information:
- Did any of their strategies work in other worlds?
- Did they work for the same reasons as they did in this world or for other reasons?
- Were there some that worked only in this world?

- **Creating a Strategy Matrix**

  Donna gathers GGF members to build the strategy matrix. The matrix arrays all the strategies in the rows and the scenarios in the columns. The ambassadors take turns reporting whether each of their strategies worked [+], didn't work [-] or didn't particularly help or hurt [~] and why. GGF conducts a systematic analysis and evaluation of each strategy.
  - If a group reports that a strategy works in a scenario other than the one it was developed for, they probe deeply into how and why it did so. Often they find that it works for an entirely different reason than it did in its original scenario. GGF traces the impact of various strategies through the structural model to test their assumptions.
  - If a strategy doesn't work, they consider the impact it does have.

Strategies \ Scenarios	We can work it out!	Tough Times	Reboot	Quiet Power
We can work it out! Strategy 1	+			
We can work it out! Strategy 2	+			
We can work it out! Strategy 3	+			
Tough Times Strategy 1		+		
Tough Times Strategy 2		+		
Tough Times Strategy 3		+		
Reboot Strategy 1			+	
Reboot Strategy 2			+	
Reboot Strategy 3			+	
Quiet Power Strategy 1				+
Quiet Power Strategy 2				+
Quiet Power Strategy 3				+
GoGo Strategy 1				
GoGo Strategy 2				
GoGo Strategy 3				
GoGo Strategy 4				
GoGo Strategy 5				

Strategy Matrix

GoGo Case

Each cell of the matrix receives a thorough discussion and questioning. A lot is riding on the outcome of this conversation, and GGF members want to be completely sure about the viability of the strategies that they have developed, particularly those that they will be recommending.

- **Robust Strategies**
  GGF selects the strategies that they feel confident will work in any of the scenarios. These are the strategies that they will recommend for implementation. Here is a sample of the strategies that GGF found to be robust:
  - Develop strong alliances with a wide range of educational institutions.
  - Build a workforce capable of self-management while simultaneously working cohesively for the good of the organization as a whole.
  - Invest in IT tools that increase the capacity of the workforce to collaborate and allow them to function independently, as the situation requires.
  - Seek and acknowledge skills, expertise and accomplishments in people of all types.
  - Encourage continuous learning to stay abreast of changes in professional and technical fields and in GoGo's markets.
  - Elevate the formal influence of Environmental Health and Safety to enhance GoGo's partnership with the natural environment.

- **Contingent Strategies**
  The Team also selects strategies that work in several but not all the scenarios. They can hold these in reserve, in case the future emerges in a way that is close to the scenarios in which they are effective. Or, the organization could implement them, being careful to change strategies if the future unfolds in a way that does not support the contingent strategy.

- **Testing Existing Strategies**
  Using the strategy matrix, GGF looks at the viability of strategies that GoGo is pursuing or plans to implement. For example, GoGo has already made a number of investments in social networking software. After careful consideration, GGF deems this strategy to be robust. On the other hand, a proposal is presently under development to build an advanced manufacturing facility for utility vehicles in Arkansas because of the low labor rates there. This plan now seems more questionable. In "We can work it out!" and "Tough Times," GoGo could be left with a major investment in a manufacturing facility designed to produce products for which there are few customers. Furthermore, Arkansas doesn't have the diversity profile that GoGo's robust strategies emphasize.

  Based on the stress testing, GGF concludes that it makes sense for GoGo to put plans to buy real estate on hold in favor of long-term leases. The degree of uncertainty surrounding future events indicates that the company is better off keeping its real estate options open rather than making commitments at this time. Leases with five- to ten-year terms with options to buy meet GoGo's need for quality space while allowing it to change strategies in light of future conditions.

Homework: While continuing to communicate, blog and update the Wall, GGF prepares for the next Executive Briefing.

GoGo Case

◉ Executive Briefing: Options

April 13, 2010: The GGF Team holds its second briefing with executives to apprise them of its findings. Sheila, in her role as strategist, realizes that it is important for the executives to comprehend the process that led GGF to recommend these strategies. This conversation could affect the direction of the organization going forward. Adriana describes how GGF considered how the structural diagrams – developed around each of the critical uncertainties – interact with one another. GGF looked for variables that appeared in two or more diagrams. A change in those variables would have an impact on other variables in other diagrams, eventually affecting the whole system. In this way, they identified a network of interconnections among the critical uncertainties. The resulting diagram is the Structural Dynamics model. Jonathan then discusses GGF's work in groups to really get a feeling for life and work in each of the scenarios. Representatives from each of the scenario groups recreate the scenarios for the GGEC in a series of brief, compelling video presentations.

Sheila then presents the strategy matrix and takes her colleagues through the implications. The GGEC fully vets GGF's thinking on each recommended strategy. They focus hard on GoGo's existing strategies. A number of these executives have put a lot of work into those initiatives. Several of the existing strategies fall into the robust category; building a plant in Arkansas is contingent.

Once the executives are satisfied that they thoroughly understand the thought process and the implications of the strategies, the GGEC needs time to review the findings and make decisions.

A week later, GGF is delighted to hear that the executives have endorsed the robust strategies. They do intend to continue to explore prospects for a facility in Arkansas, paying close attention to the composition of the labor pool there, the products that can be produced in the facility and the financial structure of the project [e.g., sale/leaseback, joint venture, etc.]. This Executive Briefing has been another step in the process of co-creating the GoGo of the future.

GoGo Case

Embodying Actions

🌑 Role of Leaders: Integrator

At this juncture, every member of GGF has become an integrative leader. Each one is excited about the prospect of helping to set a strategic direction to move GoGo into its future. The GGF Team has established a reputation for transparency and thoughtfulness that it now wants to use to ignite positive energy throughout the organization. GGF and the GGEC are committed to a life-sustaining path. They want all parts of the organization to share their enthusiasm and commitment.

The initiative has involved GoGo employees from the beginning. There are no surprises. GGF has incorporated employee feedback and has kept employees throughout GoGo abreast of decisions regarding robust and contingent strategies through their leaders, the Playbook and internal emails and newsletters. GGF members have engaged their colleagues, consulting with them, communicating insights, testing ideas and gathering input. They have blogged about their work and built the GoGo Playbook.

All key documents and strategic decisions are part of the Playbook, including:
- ✓ The names, biographies and video clips of all participants
- ✓ The Structural Dynamics model, including all the structural diagrams
- ✓ Blogged commentary from people in the organization, both participants and non-participants
- ✓ Scenarios in their narrative and video form
- ✓ The strategy matrix with the reasons behind each +, – and ~
- ✓ Robust strategies
- ✓ Contingent strategies
- ✓ Timetable for moving forward

GoGo Case

🌐 Inspiring Commitment

Session 6 – April 27, 2010: GGF is unanimous in saying that the intervals between meetings have been extremely valuable. They don't feel that the process would have been as well served if they had tried to push it along faster. They needed that time to absorb the learnings, test them, communicate them and let new insights germinate.

- **Preparing for the Strategic Conversation**

    As an international operation, GGF knows that it will not be possible for GoGo to bring the entire organization together in the same room; it must rely on distance communications. Some members of the Executive Committee will appear by way of prerecorded messages while others will be present. Prescott, Denise, Sheila and Donna will be physically present. Members of both the GGEC and GGF will be located in various geographic locations. The event will be coordinated through simulcasting and other technology support services. The event organizers want to capture the conversation on video so that people will later be able to access all the information and ideas.

    GGF decides to hold the strategic conversation in two parts, starting with the whole organization and then allowing time for the various parts of the organization to talk among themselves.

    > The Organizational Conversation will be two and a half hours long; it will be held in the morning [Central Time] and will focus on GoGo's strengths and the thought process that led to the strategies.

    > The Departmental Conversations will be three to four hours in length, facilitated locally at each GoGo site, to allow the various parts of GoGo to incorporate this thinking into their own operations. The line leaders of the various components of the organization take the lead in this part of the event. People will spend most of their time working within their own groups. At various points throughout the session, the entirety of GoGo will reconvene to share thoughts from the work of the groups.

    In most places, the Departmental Conversations will follow after lunch on the same day. In Europe, they will be held the next morning to accommodate the time differences. Other geographic locations will make appropriate scheduling arrangements.

    These strategic conversations will stretch the skills and capacity of the IT department and is seen as a harbinger of how the organization may be managing globally in the future.

Homework: Once the dates are finalized, GGF swings into action. They must attend to many details to make this day a great experience for the participants.

GoGo Case

- **The Organizational Conversation: Appreciating Competencies**
  <u>May 18, 2010:</u> Part 1 of the strategic conversation begins with Denise making a brief statement about why GoGo wants and needs to be a life-sustaining organization, what that would mean to her, the people of GoGo and the company. Here's an excerpt:

  > "Even though it's become something of a cliche for organizations to talk about people as their most important asset or how they are only as good as their people, this initiative to make GoGo a life-sustaining organization is really critical to us and to our future. When my father founded this company, he did a good job of anticipating the trends of his era, like the fact that people would be living longer, healthier lives and that golf would become a popular business and social sport. He made some good decisions and pulled together fine people who worked together to create an organization that has made a positive difference in the world. They have given us this legacy to nurture, protect and develop.

  > "Now, our company stands at the beginning of a new era. Every day, we experience the impact of living in a vastly more integrated world economy, and every day we are more aware of our delicate relationship with nature.

  > "We must respond to the dynamics of our times. We've got to recruit and support people who can really grapple with complexity and yet pull together as a team moving in a specific direction. We need to be an organization that is known to be a really great place to work. We need to anticipate what will inspire the people of this organization, both right now and in the future. I am asking for your help to develop GoGo into powerful life-sustaining organization.

  > "Today, you will hear stimulating, worrisome and fascinating ideas about what the future might hold for GoGo. You've seen the Playbook, you've had conversations and you've shared your thoughts about GoGo's strategic direction. Today is a milestone. We're here to think through the actions we need to take to move GoGo into the future we envision for ourselves."

  Each organizational unit then considers a time it was *achieving its highest purpose* as an integral part of GoGo. IT group members surprise themselves when they recognize that this meeting is one of their best times as a group because they feel like a strong, capable community, with everyone supporting each other. This appreciative inquiry [Cooperrider 2008, Vogt 2008] reminds people what they are capable of achieving and primes them for the in-depth conversations to follow.

  Next, the scenario videos are presented to the entire organization. Although most of the participants are familiar with the ideas, this is the first time most of them see the videos, and they have a powerful effect. GGF members are featured in the videos, either as narrators or actors, explaining how forces in GoGo's external environment led to the world depicted in the scenario. Seeing the scenarios represented in a powerful visual form makes a big impression. Following the presentation of the four scenario worlds, Sheila moderates a question and answer period with questions coming from those present as well as those participating in remote locations.

Prescott then introduces the strategy matrix and calls upon GGF members to discuss each of the robust and contingent strategies. For example, Kim presents the robust strategy of *acting in partnership with the natural environment* and explains why it works well in each of the scenarios. Jazmin presents the contingent strategy of *investing in centralized, owned facilities* and explains how it works in Status Quo but could become a liability in the other scenarios. They take time to walk through these explanations and field clarifying questions. Because many people have already seen and studied the strategies and the thinking behind them in the Playbook, the discussion is lively and moves along expeditiously.

Wrapping up Part 1, several employees who represent certain demographics [e.g., parents with small children, employees working part time after retirement age, workers caring for elderly parents, remote workers feeling isolated, etc.] have been selected to discuss what it would mean to them for GoGo to become a life-sustaining organization. In preparation, they have been moderating conversations by way of the Playbook with colleagues in similar situations.

- **The Departmental Conversations: Embedding New Approaches**
Part 2 begins after lunch in St. Louis and in other places where the time zones make this feasible. Each of the component groups within GoGo work on their own. They begin by articulating what the robust strategies mean to them and then brainstorm actions they can take to reflect these strategies within their own organizations. Several individuals speak about what they are going to do personally to embody the strategy. The groups then spend time considering the contingent strategies and the conditions that would cause them to embrace [or abandon] them.

Sheila reconvenes the companywide conversation to see how the groups are doing. She asks several functions and business units to describe how they plan to embrace the strategies. The Belarus plant reports that it is going to become much more active in reaching out to environmentalists in that part of the world to see how to increase efficiency, lower cost and bring innovations to GoGo's operations and distribution strategies. This strategy would seem to work across all scenarios, including Collapse, because having efficient, low-impact operations would help GoGo get through the worst of times. These brief presentations increase the positive momentum of the event.

The various component parts of the organization return to their individual conversations, this time to set dates and responsibilities for the action items they have identified.

When Event Session 2 is complete, each group gets a transcript of their conversation mapped to the video recording, so that every interaction can be located very quickly in the video record to enhance the learning going on in that part of the system.[30] These transcripts are provided only for the use of the groups that made them [Tom, SVP of HR, felt strongly that any other use of the material would create a sense of oversight inconsistent with the creation of a Life-Sustaining organization].

---

[30] The *New York Times* and other institutions have been using "flash transcription" services

G<u>o</u>G<u>o</u> Case

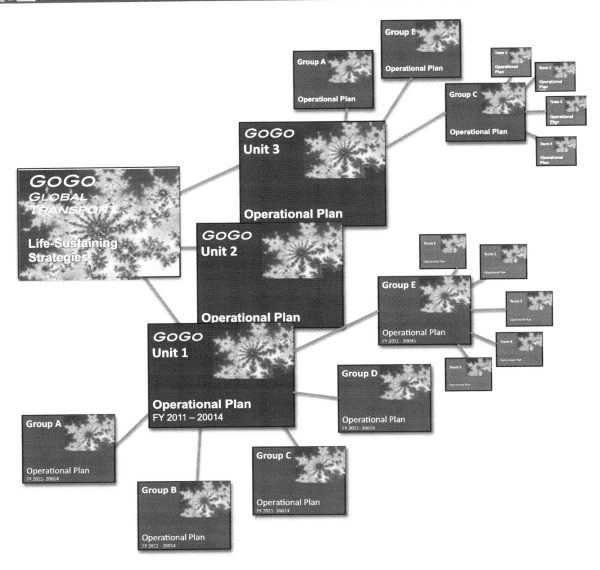

◎ Taking Action

Session 7 – June 8, 2010: By this date, three weeks after the last session, each group has posted its operational plans in the Playbook. Because of the fractal nature of the process, each part of the organization understands how it relates to and supports the company as a whole.

By clicking on the boxes in the diagram in the Playbook, every GoGo employee can view the operational plan for any part of the organization. In this way, the GGEC and anyone in related groups is able to see how various parts of the organization are embodying the strategies to become a life-sustaining organization. The organizational components derive ideas from one another that might apply in their own part of the business.

Some groups have chosen to enhance their plans with video and other innovative means of communication. As a result, these video playbook entries can be quite lively and compelling. For example, the marketing department posted a story showing an island off the coast of Massachusetts that used golf carts exclusively for transportation – saving on carbon, reducing noise and making the streets pedestrian friendly. [Note: This video later led to an ad campaign to promote the idea in golf resorts, retirement communities and even parts of historic towns.]

Today, GGF meets to analyze the various plans, sorting the strategies into themes and categories. They find that several units plan to make special efforts to recruit and support women with scientific and engineering backgrounds, for example. GGF also looks for what's missing from the various plans that might be a good fit for that function.

Homework: The Team will continue populating this analysis in preparation for the Executive Briefing.

Executive Briefing: Action

June 15, 2010: The meeting begins with a review of GGF's analysis of the operational plans to see that they are significantly in alignment with GoGo's life-sustaining strategies. While the executives find the plans overall to be excellent reflections of the company's strategic intent, they agree with GGF that several could be strengthened, improved or refocused. The appropriate member of the GGEC will work with his or her groups to improve the alignment. The GGEC and GGF then review, at a high level, the annual budget allocations needed to support the plans. The conversation then turns to an exploration of ideas for monitoring the results of the strategic initiatives worldwide and sustaining GoGo's overall commitment to the creation of itself as a life-sustaining organization. This is where GGF will focus its attention in the upcoming, final weeks of the process.

GoGo Case

Sustaining Results

🔵 Anticipatory Leadership

Since GoGo began its life-sustaining organization initiative, most members of the GGF have become quite skillful in the art of Anticipatory Leadership. Their participation in this initiative has meant a great deal to each member; in particular, they have appreciated the experience of teamwork, the development of leadership skills and the opportunity to broaden their horizons within GoGo and beyond. They have grown as employees of GoGo as a whole, rather than simply within a particular function. All GGF members want to continue to be part of GoGo's strategic renewal from wherever they sit in the organization, enthusiastically supporting the embodiment of the life-sustaining strategies. Some have become more interested in [and qualified for] assuming leadership positions at GoGo. Even though not all will continue to work for GoGo, either because they are nearing retirement age or because their personal or professional plans involve a move away from the company at some point, each one recognizes and appreciates this growth experience. Each member of GGF has come to understand the importance of thinking and acting simultaneously as an integrator, a futurist and a strategist while bringing their own styles, life experiences, backgrounds and individual orientations to the challenge of Anticipatory Leadership.

- **Monitoring Results**
  Session 8 – June 29, 2010: As the annual budgeting cycle draws to a close, GGF meets for their last session to determine which signposts, indicators and warnings to watch for in the workplace and organizational environments. Based on the discussion at the Executive Briefing, GGF recommends that a variety of elements be incorporated into an dashboard to monitor changes within GoGo [signposts], the impact of these changes internally and externally [indicators] and relevant events in the external environment [warnings] that might indicate which of the scenarios best approximates the actual future as it takes form. Decisions on what to include will lead GGF to be in communication with leaders throughout the company.

  Eleanor starts by suggesting that the credit markets are great indicators of which way the future is heading. Reliably available credit, stable markets and low interest rates support economic growth and risk taking. Volatility and uncertainty, high interest rates and an unwillingness to extend credit all combine to reduce organizations' ability to invest in the future.

GoGo Case

Logan wants to find a way to measure the degree of openness and transparency in conversations up, down and across the organizational structure. He expresses concern that people may have become too sensitized to differences in a diversified workplace, inhibiting their ability to communicate openly and honestly. On the other hand, he wouldn't want to see open communication result in hurt feelings and hostility. A life-sustaining organization needs to achieve this kind of delicate balance.

Sheila suggests that they use the characteristics of a life-sustaining organization to help them organize their thoughts:

- ✓ Creative people
- ✓ Whole systems thinking
- ✓ Design integrity
- ✓ Elegant solutions
- ✓ Results orientation

Katherine adds that they should look for signposts, indicators and warnings for each characteristic. In that way, they are sure not to miss any broad areas they should be considering. GGF members like this approach and decide to try it. They determine that credit markets are a warning of economy health or volatility; they could affect the company's ability to invest in creative talent, design integrity and elegant solutions. Open communications, they decide, is a signpost relating to whole systems thinking.

They continue in this way until they have covered the full spectrum of combinations between the signposts, indicators and warnings associated with the five elements of living systems.

Homework: Review the reports coming into the Playbook from all parts of the organization. Organize the information on the early results of the life-sustaining initiative in preparation for the final Executive Briefing. GGF members also continue communicating, blogging and updating the Wall.

GoGo Case

July 1, 2010: GoGo's annual budget is complete. Allocations are tied directly to the approved operational plans. The interactive diagram in the Playbook looks like this:

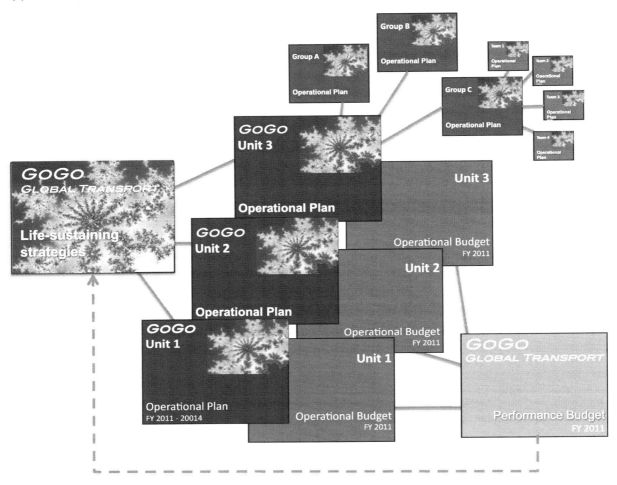

This graphic demonstrates how GoGo's strategies at every level link directly to the annual operational planning and budgeting cycle. The whole system is interconnected seamlessly in support of moving GoGo toward its objective of becoming life-sustaining.

Note: The feedback loop at the bottom of the diagram indicates that the performance budget, by comparing actual spending to plan, provides information to the GGEC pertaining to the implementation of the strategic plan.

GoGo Case

🌐 Executive Briefing: Results

<u>July 27, 2010</u>: The results of the initiative are just beginning to emerge. Employees are posting reports in the Playbook about tangible steps being taken toward become a life-sustaining organization. GGF has organized the information and reviews it with the GGEC:

1. <u>Creative Talent</u>
   Relating to this aspect of being life-sustaining:
   - GoGo employees are blogging on popular social networking sites about working at GoGo, its green products and the life-sustaining initiative. This is creating a buzz about the company.
   - GoGo's reputation as a "green company" is getting attention in the newsletters of professional schools.
   - Denise and Prescott Chaffrey appeared together on CNBC to discuss GoGo's approach to managing people.
   - Based on this appearance, they have been invited by several organizations to speak at conferences on topics ranging from "creating and sustaining a diverse workforce" to "companies potential involvement in revitalizing the communities in which they operate."
   - GoGo has been mentioned favorably in articles in *Business Week, The Wall Street Journal* and *The New York Times* on subjects ranging from leadership style, remote work and low-impact business operations.
   - The positive press is increasing GoGo's cachet with the creative class workforce it is seeking to attract.
   - The numbers of applications for open positions are increasing. Some applicants mention GoGo's "life-sustaining" initiative and reputation as a factor that attracted them to the company.
   - In the face of the changes emerging from this initiative, one line leader with a controlling style has enrolled in an Anticipatory Leadership workshop.

2. <u>Whole Systems Thinking</u>
   The integration of the operating plans with GoGo's life-sustaining strategies is affecting the organization as a whole.
   - ✓ <u>Whole Person</u>
     GoGo is becoming more focused on the knowledge, the skills, the aspirations and the concerns of each member of its workforce as a result of the life-sustaining initiative.
     - By involving a diverse cross section of employees in the initiative, GoGo demonstrated its recognition that insight, wisdom and ideas are widely disseminated throughout the organization and that it values these aspects of its employees.
     - Each employee, including the GGEC members, will develop and maintain a

GoGo Case

personal development plan with their managers.[31]

- GoGo has budgeted funds for educational opportunities for every employee regardless of their position in the company. Employees select the conferences, courses, workshops and/or programs to attend; their managers approve the choices based on the employee's development plan.
- The life-sustaining initiative encouraged employees at all levels to engage in conversations about the company and where it is going. These are no longer topics of concern to only a few people at the top. GoGo has become the business of all of its employees. There is a pervasive sense of "being in this thing together."
- Employees are tuning in to one another. There is increased sensitivity to what is driving people, what is on their minds and what their aspirations are.
- Everyone has a better sense of who has information and skills they might need to do their jobs more effectively.
- Leaders encourage ideas, creative effort and distinctive results to emerge.
- Several members of the GGEC and the GGF plan to write articles or address audiences on GoGo's model for a new kind of organization and a new way of developing strategic direction. This will increase their visibility and enhance their writing, presentation and speaking skills.

✓ Whole Organization
The units are beginning to see themselves as integral fractals of the organization.
- Recruitment choices, real estate decisions, approaches to rewards and recognition, training options, norms for communication and behavior, and partnerships with suppliers and customers have aligned in support of the life-sustaining strategies while allowing a high degree of freedom to each unit in how it manifests GoGo's strategies in action.
- Because everyone has had the opportunity to participate in the design of the life-sustaining strategies, the company has much less need for controls. People understand the right thing to do for the organization.
- GGF is considering whether the inclusive nature of the life-sustaining initiative with its emphasis on continuous learning may allow GoGo to eliminate some bureaucratic processes and reduce the number of layers in the hierarchy.
- The leadership skills that the members of the GGF developed are being emulated by others with whom they come into contact.
- GoGo doesn't need to rely on any particular individual or group to continue to advance in a life-sustaining direction.
- Because so many people are involved in strategic action, GoGo is becoming an integrated learning community. A high degree of interconnection creates nearly

---

[31] Human Resources has posted instruments, like the Myers Briggs Type Inventory, FIRO-B, the Thematic Apperception Test, the Enneagram, etc., in the Playbook for employee use as input to conversations about personal and/or professional development with their managers. These devices are seen as one way for people to know more about themselves to align their work at GoGo with their personal aspirations.

instantaneous communication of information, ideas and insights. Rapid feedback elevates good ideas and critical information.

- Increasingly disagreements and conflicts are becoming mechanisms for learning. Employees and work groups still experience some discomfort and displeasure with each other. However, the overall frame for the tension has shifted. The parties in several long-standing feuds have come to see them as systemic rather than personal issues.
- When people of different ethnic backgrounds find themselves in conflict, they are now able to view these challenges through the lens of different sets of assumptions based on cultural myths and metaphors.
- GoGo's risk managers have begun to internalize the idea that the improbable is still possible [Taleb 2007]. Thus, it is considering elegant, robust strategies to ensure business continuity in a wide range of future possibilities.
- The GGEC has decided that each of its members will have the knowledge and expertise to regenerate the firm. This is approach has profound consequences for learning and effectiveness even when continuity isn't an issue.
- GoGo's open systems approach to finding and using talent and creativity is bringing it into contact with systems thinkers around the world. Some, to their surprise, are quite nearby – such as those influenced by Buckminster Fuller's work at Southern Illinois University.

✓ Whole Planet
The life-sustaining initiative has alerted GoGo to the significant effect it could have on the well-being of the planet.

- The facilities group is exploring the cost of retrofitting existing facilities and building new ones in ways that will earn LEED[32] certification.
- GoGo has become more conscious of locating facilities in places where employees can make effective use of public transportation.
- The company plans to support and use alternative, sustainable fuels to generate electricity, realizing that most of the electricity generated in the U.S. is produced in coal-fired plants.
- GoGo has become interested in supporting the green development of St. Louis and other communities in which it is located. By making these places more attractive to employees, they will be better able to hire and retain the talent they need.

3. <u>Design Integrity</u>
GoGo has developed an appreciation for the effect of design quality on the productivity of its employees and the choices of its customers.

- The workplace group plans to involve employees in the layout and design of their workspaces. Employees will have control over how space is used as long as the work group agrees and no one is negatively affected.

---

[32] LEED [Leadership in Energy and Environmental Design] is an internationally recognized green building certification system created by the U.S. Green Building Council [USGBC]. LEED rates the design, construction and operation of buildings on energy efficiency, waste reduction and healthful quality of life.

GoGo Case

- GoGo is giving authority to its industrial designers to ensure that its products and packaging are well designed, use renewable materials and are completely recyclable.
- To reflect its commitment to its customers and employees, to the environment and to design principals, GoGo will adopt a consistent "look and feel" to convey an integrated, authentic message. It will be used in its logo, its facilities, its promotional materials and its products.

4. Elegant Solutions
The organization is seeking elegant solutions to help it make good choices among a growing number of options.
- GoGo's reputation as an employer of choice is likely to grow. Human Resources plans to screen applications with what it calls *Renaissance Parameters*:
  - Disciplinary depth, i.e., a demonstrated expertise or proven aptitude in some domain of importance to the company
  - Multidimensional attention, i.e., being curious across a number of domains
  - Authentic listening, i.e., exhibiting respect for what others say, even when the view expressed is not one with which the listener agrees
- A workplace team composed of leaders from personnel, facilities, business development, information technology and finance is coordinating the company's policies regarding working from home to ensure that they are facilitating productivity rather than creating unnecessary hurdles that may be cumbersome, stressful and time consuming.

5. Results Orientation
GoGo is on its way to becoming a deeply innovative, life-sustaining company.
- The GGEC has discovered that its employees possess a much richer set of ideas for creating products and for managing work-life issues than they had previously realized.[33]
- GoGo is becoming more inquisitive and more appreciative, seeking out ideas from many business and industrial sectors, creative thinkers and social phenomena to find ideas that it would once have considered irrelevant.
- The workplace team is tracking promising developments in the field of remote work. Members plan to invite representatives of companies with innovative approaches to a day-long meeting at GoGo. The expected results for GoGo include:
  - Access to the latest thinking and research on remote work
  - Ideas regarding the most effective technology to support remote workers and pointers toward strategic next steps
  - Possibilities for reductions in the company's real estate holdings

---

[33] Several employees recently designed a prototype for a potentially exciting new product, the Do-All-Dolly, a small piece of equipment running on advanced battery technology. Its four wheels would adjust automatically to move heavy loads up and down stairs. Several customers have expressed interest: a moving company has volunteered to test a prototype because they can envision the safety and productivity benefits it would have.

The GGEC is impressed with the amount of change that it can observe in so short a time. It wants to be sure to maintain the momentum. To that end, the company will create a small cross-functional group within the strategy function. This group will work with information technologists to create a dashboard that has been envisioned to monitor the signposts, indicators and warnings identified by GGF. Some of the GGF members will initially staff this group for a year and be responsible for training new people who will rotate in for a period of time. The GGEC decides to invite this group to provide an update on the life-sustaining initiative every six weeks at its Monday morning meeting.

Final Word

- **Taking a Deep Dive**
- **The Evolution of Work**
- **Making the Mental Shift**
- **Future Mind**

Final Word

Taking a Deep Dive

> Oh, great Googa Mooga,
>   can't you hear me talking to you.
> Just a ball of confusion.
> Oh yeah, that's what the world is today.
>
> *Ball of Confusion*
> *The Temptations*

Maybe the world has always seemed like a ball of confusion; it certainly is today. We have never been confused in quite this way before. We face precarious conditions as the twenty-first century picks up steam. Daunting threats menace us at the same time that a phenomenal array of exciting, life-positive developments in science, medicine, technology and human consciousness are appearing. Now, with more than six and a half billion people on the planet, our collective choices are having a profound impact on the balance of natural systems. According to Ecclesiastes, the Earth will abide forever; how well the human species will do is an open question. Can we continue to rely on human ingenuity and technology to save us one more time? Have we reached the apex of human evolution? Do we hear the chords of a requiem scenario, climaxing with a crashing crescendo that marks the end of human civilization? At either a conscious or an unconscious level, we are aware that the stresses being placed on Nature could culminate in this horrific prospect.

Why expend the effort to create life-sustaining organizations in the face of so much uncertainty? The good news is that humanity has repeatedly demonstrated the capacity to keep moving and hoping in the face of overwhelming odds against us. Life-sustaining organizations create a rich, fertile base from which talented people do great things. Organizational systems mold our choices as individuals, groups and societies to a much greater degree than most of us recognize. Therefore, consciously fashioning our organizations to affirm life profoundly impacts the behavior of the people who embody the purpose and efforts of those organizations.

The process we've described, Structural Dynamics, can help organizations make sense of the complexity they are facing. When we base our actions solely on observation of current events, we never "connect the dots." We look exclusively at what's happening and wonder where "that" came from. Exploring a little deeper to see how these events have been trending over time, we find patterns that occur over and over again. Going yet deeper, we find interconnections that comprise whole systems: people in the context of an organizational system, an organization in the context of social, economic, communication, transportation and natural systems, and the Earth in the context of the universe. Going inward, we become aware of our personal and collective relationship to the wholeness of the reality we see – our beliefs, our mental models and our memes.

As we reach this deeper level of knowing, the source of our actions changes from reactive emotionality and short-term thinking to a stance that serves the larger whole; we begin to recognize the future that needs to emerge. We act less from self-interest and more from community; less from concentrating on the parts to optimizing the whole. We act less from competitiveness and more toward compassionate interdependency. Really knowing that we are all in this together – that we can't save ourselves or our institutions without saving the natural environment in which we are embedded – entirely changes the impetus of our actions. New insights and intuition create the possibility to act rather than react; to set a course rather than being surprised and shocked. In doing so, we shape our personal, organizational and global environments. Rather than constructing defenses to fend off what

Final Word

we don't want, we can act creatively and collaboratively to bring into existence what we do want.

The Evolution of Work

> All paid jobs absorb and degrade the mind.
> *Aristotle*
> *384 BCE-322 BCE*

We believe that the desire to be productive, to innovate and to contribute to society is fundamental to human nature. That does not mean that we have always had the same approach to work, life and leisure. Distinct differences in attitude appear across time and cultures. The ancient Greeks believed there was no real gratification to be found in paid work. The leisure class enjoyed the pleasures of the arts, philosophy and sports while serfs, slaves and servants performed necessary, routine tasks.

This attitude prevailed well into the Middle Ages. Leonardo da Vinci and Michelangelo were among the first to recognize the glories of some practical work, particularly in relationship to the arts. During the French Enlightenment, Denis Diderot and Jean le Rond d'Alembert published their twenty-seven volume *Encyclopedie* to "change the common way of thinking"

> If people knew how hard I had to work to gain my mastery, it wouldn't seem wonderful at all.
> *Michelangelo Buonarroti*
> *1475-1564*

> What is more felicitous than to please ourselves in the pursuit of occupations suitable to our talents and our condition?
> *Pleasure [Ethics], Encyclopedie*
> *1751 and 1772*

through the expansion of knowledge and the development of critical modes of thought. In this major achievement, they advanced the idea of joy in activities such as baking, planting, printing and operating an enterprise. The pursuit of satisfaction could now be found in work rather than leisure. Activities with no financial reward were of diminished importance [de Botton 2009].

The nineteenth century ushered in the Industrial Revolution and with it some amazing innovations and significant progress: life expectancy increased, literacy rates and human rights expanded, and technology made giant strides. It also engendered a way of thinking about ourselves and our institutions based on mechanistic models. We broke work into incremental parts and measured efficiency. We rewarded working faster, not smarter. Workers knew their piece of the operation, not the whole. Control, predictability and standardization became valued over spontaneity and innovation. Work was separated from the rest of life as it had never been before. As part of his revitalization of Paris in the 1860s, Baron Haussmann created zones for workplaces.

In 1922, architect Le Corbusier proposed a plan for the "Contemporary City" based on efficiency concepts adopted from industrial strategies. At the center of this city was a transportation hub surrounded by huge office towers which in turn were encircled by smaller apartment blocks in park-like settings. Although this plan was ultimately abandoned, it had a significant impact on

> The proposed city appeared to some an audacious and compelling vision of a brave new world, and to others a frigid megalomaniacally scaled negation of the familiar urban ambient.
> *Norma Evenson, 1969*

181

Final Word

the subsequent design of cities worldwide.[34] Work was no longer close to home in a community where everyone knew your name and everything else about you. Workers now commuted to urban centers or industrial zones on the urban periphery.

Complex webs of rich personal and communal relationships gave way to aggregations of strangers pursuing a purpose defined by someone "up there" and engaging each other in relationships bounded by the constraints and conventions of organizational roles. Privacy became something to be scrupulously guarded.

The twentieth century witnessed the emergence of global corporations enabled by colossal financial, distribution, agricultural and telecommunications systems developed in parallel in a cycle of reinforcing massive organizational size. People became things, replaceable cogs in a machine. Employees were seen as necessities rather than assets. Loyalty between employer and employee largely vanished. Automation came to be preferred to human labor whenever possible. The bottom line ruled without any consideration of the interconnection of financial and natural systems.

> In the long term, the economy and the environment are the same thing.
> — *Mollie Beattie, Director U.S. Fish and Wildlife Service, 1996*

These institutions and their infrastructures had a profound influence in other domains, standardizing educational and healthcare systems, and spreading a culture of consumerism worldwide. Developed countries and those seeking to become developed responded to big business and shaped their political, trade, economic and environmental policies to accommodate these behemoths. These complex systems, perceiving themselves as machines, seek equilibrium – an optimal state in which they can continue to grow without interruption. Most of these multinational institutions operate without regard to their collective impact and the natural and societal consequences of their actions. They grow without any purpose other than self-preservation. This network of giants is able to pollute the air, befoul water supplies, eliminate species and change climate patterns. The unanticipated consequences of these actions are generating complex issues that risk the survival of humanity – dramatic weather events, the disruption of ocean currents, droughts resulting in food and water shortages, melting ice caps and rising sea levels.

> The shark [is]...a killing machine...There isn't any question of malevolence or of will; the enterprise has within it, and the shark has within it, those characteristics that enable it to do that for which it was designed.... [the corporation] is potentially very, very damaging to society.
> — *Robert Monks, as quoted by Drutman and Cray, 2004*

Critics believe the corporate form itself is "inherently destructive." A number of organizational theorists hold the view that large complex organizations are so broken that the form is collapsing and needs to be quarantined and discontinued. We haven't given up on complex systems and global enterprises. In fact, we believe they are critical to the world's health and survival.

---

[34] Efforts to eliminate slums in the 1950s used this model for low-income housing towers. Many of these complexes survive, although most have been recognized as planning disasters and some have been demolished.

Final Word

David Cooperrider, the father of Appreciative Inquiry, writes that "Business – with the most adaptable organizational forms ever invented and its agility, connective technologies, and its penchant for pragmatic entrepreneurship and continuous learning – could contribute to the well-being of many." He asks, "How might we...meet one another across civilizations, cultures, belief systems, nations, traditions, and worldviews not with frozen positions and answers but with creative questions, a deep and sincere spirit of inquiry and an openness to discovery, surprise, new knowledge, and a sense of awe?" He yearns for "world-centric dialogue" about business as a positive force in lieu of "polarizing ethnocentric debate."

> "We need to spread as widely as possible the image of business as one of the great creative forces on the planet...It is rapidly becoming clear what is not working; we have yet to form a vision of the global society that does work...a task of historic proportions."
>
> *Willis Harman, as quoted by Cooperrider, 2007*

> The basic problem with the new species of global institutions is that they have not yet become aware of themselves as living.
>
> *Peter Senge, 2004*

Increasingly, organizations are emerging that know, at a deep level, that they are living organisms intricately connected to nature. Gretchen Daily, professor of biology at Stanford and one of the founders of the Natural Capital Project, receives copious requests from corporations and other organizations seeking to harmonize organizational activities with natural processes [Kleiner 2009a]. Organizations can learn and evolve as they grow, with their people acting as a highly attuned, networked nervous system. Life-sustaining organizations consciously embrace change, replacing the desire for stable equilibrium with a sense of discovery and adventure, as "part of an ever-changing, interlocking, nonlinear, kaleidoscopic world" [Brian Arthur as quoted by Waldrop 1992].

These are the organizations that are elaborating new dimensions of possibility. Working close to home is again an option due to the declining cost of communication and other technological innovations combined with new styles of organizing and managing. People are living longer and healthier due to medical advances and preventive measures. We are working longer both out of financial need as well as a desire to remain active and engaged in the community and the world. As a consequence, many more generations are contributing to the workforce simultaneously. Gender, racial and ethnic diversity are also more commonplace, adding a rich mix of perspectives. A multiplicity of new doors are opening into the work environment.

> Sooner or later I'm going to die, but I'm not going to retire.
>
> *Margaret Mead*
> *1901-1978*

Life-sustaining organizations are aware of their interconnectivity with the domains in which they operate – natural, social and economic. These living systems exert great leverage on the definition of what work is and the expectations of the work environment. They are the leaders. These institutions are awakening to their responsibilities to safeguard the natural environment for the survival and well-being of the human species. They realize they can no longer despoil a corner of Earth and then move on; boundless frontiers exist only in outer space. Life-sustaining organizations increasingly act in ways that reflect the consequence of knowing that the reality of life on this planet is an intricate web of natural laws and human choices.

Final Word

 Making the Mental Shift

For nearly two centuries, we have been viewing our organizations as mechanistic constructs with replaceable parts and uninspiring goals. To move to the realization that they are living systems capable of sustaining the life of their workforce, themselves and the global environment requires a basic shift in what and how we think.

> To understand some phenomenon or set of phenomena, first rid your mind of all preconceptions.
>
> *Rene Descartes*
> *1596 – 1650*

We tend to rely on cultural, patriotic or religious mental models, myths and metaphors as a short hand to relate new experiences to something we already know. They provide a sense of order; they keep us psychologically comfortable and secure. The problem is that they may be out of step with what is actually happening now, in this situation.

As we shift our thinking from presumption and unarticulated theories to a grounded awareness of what really exists, we stop relying on beliefs formed from prior impressions, reactive reasoning and historic learning. We turn instead toward deep observation, direct inquiry and profound listening. We move from assuming that we should know the answer to assuming we don't. We look at information and situations with a fresh mind. When we open ourselves to investigation, insight, intuition and inspiration, we are comfortable living in the question. We realize that our minds have a causal connection to the world as we experience it. Fresh, original thoughts are far more powerful than habit-based, conventionally determined conclusions.

Of course, questioning everything all the time would be extremely burdensome and time consuming. We have to be able to discern when to slow down and consider more deeply. We can do so by developing the mental agility to be open to the possible. When we develop the capacity to remain open, to be vulnerable, to become mystified and to be unsure, our thinking starts to shift. We ask questions and listen deeply to what emerges. We let go of what we think we know to allow in "what is." We value time spent in the question, in the not-knowing. If we find ourselves making comparisons to past experiences, we ask, "Is it really like this?" rather than simply moving forward. Fresh thoughts, insights and creative ideas emerge from deep listening and observation. We put our old tapes on pause to make mental room for what is real in the present. In this state of deep listening, our experience is unfiltered. We become capable of seeing the familiar in new ways – directly and authentically, free of outdated mental baggage.

This shift in thinking allows us to see ourselves, our organizations and nature as living, dynamic systems in motion. The practice of inquiry and openness leads to a deeper understanding of reality. As our thinking penetrates more deeply, we see a larger perspective, perceive the whole organism and sense our own place in that wholeness. When this happens, we act from deep knowledge and understanding. We see the world in a new way. We become fresh, ready and eager to support new possibilities.

Seeing our organizations as alive and life-sustaining provides us with an opportunity to take an active role in the evolution of our institutions: to act in the world as it is and to co-create

Final Word

the world we want. The power and effectiveness of our actions is greatly amplified by knowing where, when and how to nudge the systems we care about.

Our unconsidered mental models and fragmented thought processes have taken us to the edge of the precipice, dangerously close to the Collapse scenario. Understanding our organizations as living systems that are truly life-sustaining rather than life-draining offers the possibility of moving toward generative solutions, a New Reality scenario in which all life flourishes. If we foresee the range of possibilities and experience their potential reality, perhaps we will be galvanized to take action to avoid Collapse, before it is too late.

> Unanswered questions are far less dangerous than unquestioned answers.
>
> *source unknown*

Future Mind

Life-sustaining organizations hold an expansive view of the future – a view that engages
    the humanity of individuals,
        the health of communities,
            the highest purpose of themselves as institutions, and
                a passionate desire to be in harmony with the Earth.

We call this way of thinking, sensing and knowing "future mind."

➡ The future mind is open to knowledge and excited about the co-evolution of humanity, science and nature.
➡ It has the discipline, imagination and creativity to respect the uniqueness of every moment.
➡ Those with future mind are frequently meditators, inquisitive about the interior dimensions of the mind's ecology.
➡ They find neuroscience and spirituality fascinating.

Leaders with future mind synthesize disparate ideas, into a complex, vibrant whole. Their insights into the intricate and delicate dynamics of systems incline them toward ethical behavior.

Individuals, organizations and communities with future mind are awake, in tune with complexity, able to think, willing to wait and ready to act. Theirs is the consciousness that lifts the sights of humanity from oppressive fear to a humble, yet joyous, curiosity.

> The future is literally in our hands to mold as we like, but we cannot wait until tomorrow. Tomorrow is now.
>
> *Eleanor Roosevelt*
> *Human Rights Activist*
> *1884 – 1962*

Structural Dynamics
Strategic Leadership Process

Process Overview

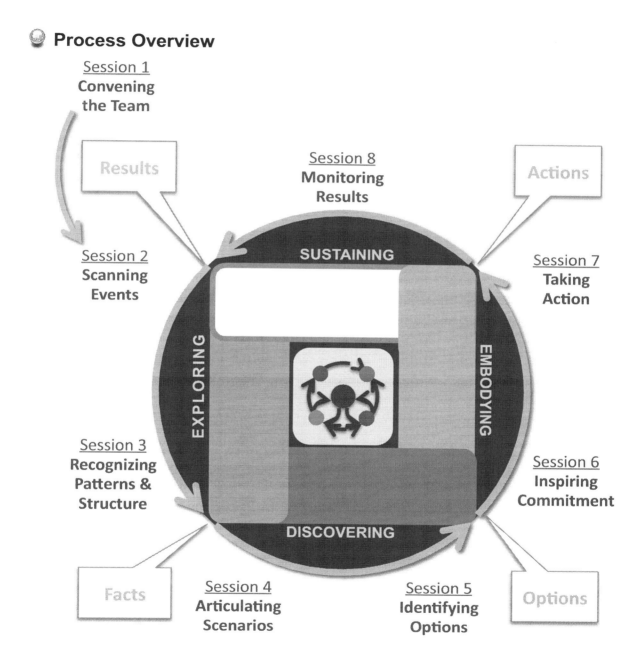

Structural Dynamics

💡 Sessions in the Process

Session 1 – **Convening the Team**
Team members learn more about each other and the goals of the initiative. They identify the strengths of the organization and define the Decision Issue, a time horizon, and criteria for documentation and communication.

Exploring Facts

Session 2 – **Scanning Events**
The Team explores for events relevant to the Decision Issue amid a vast amount of information. They focus on the most critical and the most uncertain. They recognize how these seemingly disparate events are interconnected and interdependent.

Session 3 – **Recognizing Patterns & Structure**
A pattern is a set of durable connections among related events. A system of interacting patterns forms a structure. Understanding how this structure drives events provides insights about how the future might unfold.

> Executive Briefing – **Facts**
> After exploring the external environment, the Team discusses the facts they have found with the decision makers. This strategic thinking provides a solid foundation of agreement for the strategic inquiry.

Discovering Options

Session 4 – **Articulating Scenarios**
Using the facts depicted in the model of the structure, the Team identifies a set of divergent future scenarios related to the Decision Issue. All of these scenarios are plausible because the model reveals how they could evolve from the present.

Session 5 – **Identifying Options**
The Team "lives" in the scenarios and develops options to address the Decision Issue. Testing these options across the scenarios, members discover which work in any future and which could most effectively influence the course of future events.

> Executive Briefing – **Options**
> Having discovered robust and contingent strategic options, the Team reviews these with the decision makers. The most robust options become strategies that allow the organization to take immediate steps to move into the future with confidence.

Structural Dynamics

Embodying Action

Session 6 – Inspiring Commitment
To move from strategic analysis to strategic action, each component of the organization and every member of the workforce needs to embody the strategies and apply them within their own areas of responsibility.

The Strategic Conversation
The executives sponsor an event that includes everyone in the organization physically or virtually, to communicate their support. The various functions and units develop and share implementation ideas for their areas of responsibility.

Session 7 – Taking Action
The organizational components make plans to operationalize the strategies. Analyzing these plans, the Team identifies themes and communicates creative concepts broadly across the organization.

Executive Briefing – Action
As the organization embodies the strategies, the Team reviews its analysis with the executives to assure alignment of the operational plans throughout the organization.

Sustaining Results

Session 8 – Monitoring Results
Team members determine which signposts, indicators and warnings to watch for in the workplace and organizational environments to determine the results of the organization's strategic actions, internally and externally.

Executive Briefing – Results
The results of the initiative are beginning to emerge. The Team reviews its monitoring approach with the decision makers. The executives decide how to sustain the learning and accelerate the forward momentum.

Structural Dynamics

Scenario Game Board ©2010

The Heart of Structural Dynamics

- Economic depression
- Socio-political disruption
- Technological decline
- Survival mentality
- Anxiety & cynicism

- Economic prosperity
- Socio-political transformation
- Tech & genomic breakthroughs
- Radical new forms of belief, behavior & organization

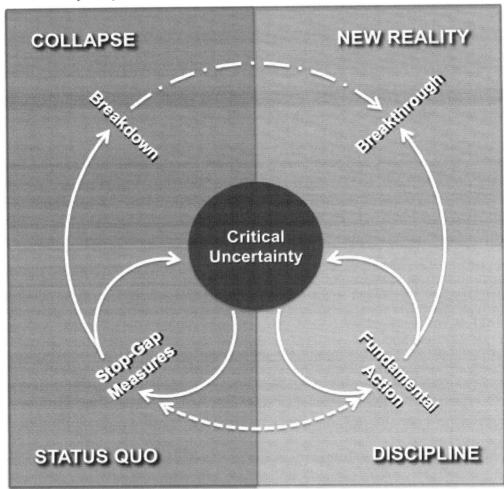

- Economic disparity
- Socio-political nostalgia
- Technological maturity
- Fear of change
- Traditional forms of belief, behavior & organization

- Economic investment in future
- Socio-political idealism
- Technological innovation
- Willingness to sacrifice
- Emerging new forms of belief, behavior & organization

Influences and Resources

It would be nearly impossible to identify all the resources we've drawn upon to write this Guide. The following is a sample of those works that have influenced our thinking and bibliographic citations for those we have referenced in the Guide along with some commentary.

Alexander, Christopher [2004], *The Phenomenon of Life: Nature of Order, Book 2: The Process of Creating Life.* Berkeley, CA: The Center for Environmental Structure [design integrity, systems thinking, elegant solutions]

_____ [2002], *The Phenomenon of Life: Nature of Order, Book 1: An Essay on the Art of Building and the Nature of the Universe [The Nature of Order]*, Berkeley, CA: The Center for Environmental Structure [design integrity, sense of place, elegant solutions]

_____ [1979], *The Timeless Way of Building,* Oxford: Oxford University Press [design integrity, sense of place, systems thinking, elegant solutions]

Alexander, Christopher et al. [2004], Gallery "Toward a New Architecture of Life," www.katarxis3.com/Gallery.htm [design integrity, systems thinking, living systems, elegant solutions]

Alexander, Christopher, Ishikawa, Sara and Silverstein, Murray [1977], *A Pattern Language,* Oxford: Oxford University Press,. [design integrity, sense of place, systems thinking]

Andrews, Michelle [2009], "In the Future; Doctor Shortage Is Projected," *New York Times,* 9/6/09. Powerful graphs showing huge shortfall in the number of physicians needed in the US. [creative talent]

Anklam, Patti [2007], "Net Work: A Practical Guide to Creating and Sustaining Networks at Work and in the World," Oxford, UK: Elsevier [workplace dynamics]

Argyris, Chris [1990],*Overcoming Organizational Defenses. Facilitating organizational learning*, Boston: Allyn and Bacon [systems thinking, organizational learning]

_____ [1964], *Integrating the individual and the organization.* New York: Wiley & Sons [workplace dynamics, anticipatory leadership]

_____ [1957], *Personality and organization; the conflict between system and the individual.* Oxford, England: Harpers [workplace dynamics, systems thinking]

Argyris, Chris and Schön, Donald [1995], *Organizational Learning II: Theory, Method, and Practice*, Reading, MA: Addison-Wesley Longman [systems thinking, organizational learning]

_____ [1978], *Organizational Learning: A theory of action perspective,* Reading, MA: Addison-Wesley Publishing Company. [Topics, systems thinking, organizational learning, workplace dynamics]

Argyris, C.; Putnam, R.; and Smith, D. [1985], *Strategy, Change and Defensive Routines.* San Francisco: Jossey-Bass [whole systems thinking, organizational learning, workplace dynamics]

Armas, Genaro C. [2001], "By the numbers, population gets older," *The Boston Globe,* 12/13/01. Those over age 65 have tripled since 1950. [generational dynamics]

_____ [2001], "US survey shows rise in foreign-born," *The Boston Globe,* 1/3/01. "About 87% of US-born residents graduated from high school compared with 76% of naturalized foreign-born residents and nearly 60% of those without citizenship. [immigration]

Barnsley, Michael [1993], *Fractals Everywhere,* San Francisco: Morgan Kaufmann [design Integrity, elegant solutions]

Barret, Craig [2006], "Why America Needs to Open Its Doors Wide to Foreign Talent," *Financial Times,* 1/31/06 [immigration, recruitment and retention, creative talent]

Bell, Gordon and Gemmell, Jim [2009], *Total Recall: How the E-memory Revolution will Change Everything,* New York: Dutton. Remember everything and find everything that you've remembered: "Seeing patterns in the vista of your life;...travel anywhere, anytime;... maintain complete access to every detail concerning your enterprise;...transfer...memories from one occupant of a position to the next." [workplace technology, workplace dynamics]

Influences and Resources

Bennett, Drake [2009], "Thinking literally: The surprising ways in which metaphors shape your life," *Boston Globe,* 9/27/09. Metaphors are creations based on bodily sensations and affected by the body's experience. [mental models, myths, and metaphors]

Berger, Warren [1999], "Lost in Space," *Wired,* 2/99. No offices, no desks, no personal equipment. And no survivors! Report on an experimental leader. [remote work, anticipatory leadership, real estate strategy]

Berringer, Felicity [2009], "White Roofs Catch On as Energy Cost Cutters," *New York Times*, 7/29/09: Another example of how little it would actually take to have a huge impact on the so-called "energy crisis." [ecology, design integrity, systems thinking]

Beschloss, Michael [2009], "Missile Defense," *New York Times Book Review,* 10/4/09. "The survival of the United States depends upon...the ability of the Federal government to attract extraordinary talent at all levels." [creative talent, recruitment and retention]

"The Best vs. The Rest," *Working Mother*, 2008. Numerous large workplaces are parent-friendly. Covers workplace support systems [e.g., paid paternity leaves by 75% of the Best vs. 13% nationally] at specific companies. [gender issues]

"Best Places to Launch a Career," *Business Week,* 9/14/09: Includes a table listing the attributes of the Top 50 best places, e.g., Johnson & Johnson, #41, picks up full tab for graduate degrees. [creative talent]

Bishop, Peter, Hines, Andy and Collins, Terry [2007], "The current state of scenario development: an overview of techniques," *foresight*, Volume 9, Number 1, pages 5 - 25. Scenario analysis is widely used in futures studies. This comprehensive overview compares scenario techniques and comments on their utility, strengths and weaknesses. It defines, describes and evaluates eight scenario development categories along with 23 variations. The study finds that a scenario technique grounded in causal analysis of dynamic complexity "creates the best quantitative representation of continuous variables that describe the future state." This sounds like Structural Dynamics to us. [methods, scenarios]

Bloom, Paul [2009], "Natural Happiness: The self-centered case for environmentalism," *New York Times Magazine,* 4/19/09. There is a considerable mismatch between the world in which our minds evolved and our current existence. [ecology, systems thinking]

Bolman, Lee and Deal, Terrence [2008], *Reframing Organizations: Artistry, Choice and Leadership,* San Francisco: Jossey-Bass [systems thinking, organizational learning, workplace dynamics, anticipatory leadership]

Bortof, Henri [1996], *The Wholeness of Nature : Goethe's Way Toward a Science of Conscious Participation in Nature*, Great Barrington: Lindisfarne Books [whole systems thinking, design integrity]

Bradshaw, Della [2009], "Applicants flock to the Global village," *Financial Times,* 10/19/09. Two, three, four continent training MBA programs costing $100K+ are the only ones showing growth in the recent past. [creative talent, globalization]

Bradsher, Keith [2009], "China Races Ahead of U.S. In Drive to Go Solar," *New York Times*, 8/25/09. Renewable energy is popping outside the US. Suntech's massive PV-covered facility is pictured in the article. [scenarios, living systems, ecology]

Broad, William J. [2009], "In hot pursuit of fusion [or folly]," *New York Times,* 5/26/09. Fusion science has been thirty years in development with, perhaps, another thirty years of continued work before conclusions are reached. [creative talent, ecology, scenarios]

Brown, Shona L. and Eisenhardt, Kathleen M. [1998], *Competing on the Edge: Strategy as Structured Chaos,* Cambridge, MA: Harvard Business School Press [workplace dynamics, new sciences]

Brown, Tim [2008], "Design Thinking," *Harvard Business Review,* 6/08. "Thinking like a designer can transform the way you develop products, services and processes--even strategy." [design integrity, elegant solutions, living systems]

Influences and Resources

_____ [2008], *Creativity and Play,* TED presentation, 5/08. Designer and IDEO CEO, Tim Brown talks about the powerful relationship between creative thinking and play -- with many examples you can try at home, and one that you shouldn't. [design integrity, elegant solutions, living systems]

Brule, Tyler [2009], "The city of your dreams," *Financial Times,* 6/12/09. The twenty-five most livable cities, with Zurich on top: [creative talent, urban design, sense of place, real estate strategy]

"Building Better Wall Street Leaders," *Washington Post,* 9/19/09. Why high velocity change makes scenarios ideal for Wall Street and what happens to the leaders who use them. [scenarios, anticipatory leadership]

Burns, Judith [2009], "Google trick tracks extinctions," *BBC News,* 9/4/09. Scientists say Google's PageRank could be a simple way of working out which extinctions would lead to ecosystem collapse.[scenarios, elegant solutions]

Butterfield, Fox [1997], "Report Links Crime to States with Weak Gun Controls," *New York Times,* 4/9/97. Violence is an issue in the workplace and could influence where organizations locate. [sense of place, creative class, workplace dynamics]

Cahill, Thomas [1995], *How the Irish Saved Civilization: The Untold Story of Ireland's Heroic Role from the Fall of Rome to the Rise of Medieval Europe,* New York: Nan A. Talese [scenarios, living systems, organizational learning]

"California's water wars: Of farms, folks and fish," *The Economist,* 10/27/09. Power and systems story about water, infrastructure and California's extreme vulnerability. [ecology, systems thinking, scenarios]

Cândido, Carlos and Santos, Sérgio [2008], "Strategy Implementation – What is the Failure Rate?" *Associação de Tecnologias e Sistemas de Informação Estratégicos [Society for Industrial & Organizational Psychology of Portugal].* [strategic planning]

Canton, James [2006], *The Extreme Future: The Top Trends that will reshape the world for the next 5,10, and 20 years,* New York: Dutton. In the chapter on the workplace of the future, the author calls his prediction of intense talent wars "the top driver of competitive advantage." Three tables predicting top jobs over the next 20 years. [creative talent]

Caruso, David [2005], "Census says Manhattan first in single-person households," *The Boston Globe,* 9/3/05. About twenty seven million Americans lived alone in 2000, about 10% of the population. [demographics, creative talent]

Casciaro, Tiziana and Sousa Lobo, Miguel [2005], "Competent Jerks, Lovable Fools, and the Formation of Social Networks," *Harvard Business Review,* 6/05. Seminal and highly readable article on social network analysis in the workplace. [workplace dynamics, systems thinking]

Cetron, Marvin and Davies, Owen [2008], *55 Trends Shaping Tomorrow's World,* World Future Society. This publication shows how quickly things get dated, e.g., anticipating continued positive growth of the US economy and completely missing the deep recession of 2008 and 2009. The 55 trends identified contradict and cancel each other out, leaving no real clarity about how the future will unfold. That said, there is also much to recommend in this publication. It's broken down into eight broad, nested categories and offers many useful speculations/insights that can be of value to us, e.g., "All the technical knowledge we work with today will represent only 1% of the knowledge that will be available in 2050." [workplace technology, scenarios, future mind]

Cheney, Margaret [1981], *Tesla: Man Out of Time,* New York: Touchstone

Coleman, Ornette, http://en.wikipedia.org/wiki/Ornette_Coleman

Collins, Gail [2009], *When Everything Changed: the amazing journey of American women from 1960 to the present,* Boston: Little, Brown and Company [gender issues]

Conlin, Michelle [2009], "Test-Drives in the C-Suite," *Business Week,* 10/19/09. Companies are turning to tryouts to avoid hiring the wrong executives. Forty percent of all executives hired from the outside fail in their new organizations within 18 months or less! [creative talent, recruitment and retention]

Influences and Resources

Conniff, Richard [2005], *The Ape in the Corner Office: Understanding the Workplace Beast in All of Us,* New York: Crown Business [workplace dynamics]

Cook, Gareth [2002], "A Challenging view of the universe: Scientist's tome *[A New Kind of Science]* draws notice, critics," *Boston Globe,* 6/2/02. Reviews work of Stephen Wolfram, a polymath, who believes he has discovered that nature uses algorithms – cellular automata – a set of simple rules acting like computer programs that run over and over again. [elegant solutions, living systems, systems thinking, new sciences]

Cooper, Michael [2009], "In Aging Water Systems, Bigger Threats are Seen," *New York Times,* 4/19/09. [infrastructure, ecology, scenarios]

Cooper, Michael and Palmer, Griff [2009], "Cities get less stimulus for roads," *New York Times,* 7/9/09. Sprawl is being paid for by cities where it's not as prevalent. [infrastructure, urban design, real estate strategy]

Cooperrider, David [2007], "Business as an Agent of World Benefit: Awe is What Moves Us Forward," *Appreciative Inquiry Commons*, http://appreciativeinquiry.case.edu/research/. Cooperrider is the Founder and Chairman of the Center for Business as an Agent of World Benefit at Case Western Reserve University, Weatherhead School of Management. See worldbenefit.case.edu.[methods, living systems, organizational learning, workplace dynamics]

Cooperrider, David and et al. [2008], *Appreciative Inquiry Handbook.* San Francisco: Berrett-Koehler [methods, living systems, organizational learning, workplace dynamics]

Creswell, Julie [2006], "How Suite It Isn't: A Dearth of Female Bosses," *New York Times,* 12/17/06. Only about 16% of corporate officers at Fortune 500 companies are women; only 2% of CEOs are women. Extensive article. [gender issues]

Dalai Lama [2005], *The Universe in a Single Atom,* New York: Morgan Road Books [new sciences, scenarios]

Daniel, Mac [2005], "Commute delays grow as traffic shifts outward," *The Boston Globe,* 5/5/05. Los Angeles drivers spend 93 hours a year delayed in traffic [and that doesn't include the amount of time when they weren't delayed!]. Boston logs in at 51 hours a year. [urban design, infrastructure, recruitment and retention, real estate strategy]

Dator, Jim [1998], "The Future Lies Behind! Thirty years of teaching futures studies," Introduction to the special issue on "Teaching Futures Studies at the University Level," *American Behavioral Scientist.* "...all images in all cultures that I have encountered can be lumped into one of four major (generic) images of the future:
  - Continuation (usually "continued economic growth")
  - Collapse (from [usually] one of a variety of different reasons such as environmental overload and/or resource exhaustion, economic instability, moral degeneration, external or internal military attack, meteor impact, etc.)
  - Disciplined Society (in which society in the future is seen as organized around some set of overarching values or another--usually considered to be ancient, traditional, natural, ideologically-correct, or God-given.)
  - Transformational Society (usually either of a "high tech" or a "high spirit" variety, which sees the end of current forms, and the emergence of new (rather than the return to older traditional) forms of beliefs, behavior, organization and--perhaps--intelligent lifeforms" [future mind, scenarios]

Dator, Jim [2006], *Four Futures for Hawaii 2050*, Hawaii Research Center for Future Studies, Department of Political Science, University of Hawaii at Manoa, [future mind, scenarios]

Davenport, Thomas H. and Harris, Jeanne G. [2007], *Competing on Analytics: The New Science of Winning,* Boston: Harvard Business School Publishing [creative talent, recruitment and retention, new sciences]

de Botton, Alain [2009], *The Pleasures and Sorrows of Work,* New York: Pantheon [creative talent, workplace dynamics]

Influences and Resources

de Geus, Arie [1997], *The Living Company,* London: Nicholas Brealy [living systems, systems thinking]

Deal, Terrence and Kennedy, Alan [2000], *Corporate Cultures: The Rites and Rituals of Corporate Life,* New York: Perseus. [workplace dynamics, mental models, myths and metaphors]

Dedman, Bill [2002], "Proportion of married Americans drops slightly in new census," 6/5/02. "The census found 44,000,000 disabled Americans age 20 and up." How does work environment respond to these conditions? [diversity, workplace dynamics]

Deming, W. Edwards [1986], *Out of the Crisis,* Cambridge MA: MIT Press [systems thinking]

Denison, D.C. [2002], "As Work becomes decentralized, connection is crucial," *The Boston Globe,* 4/21/02. Describes a study by Michael Bell and Michael Joroff, *The Agile Workplace,* that looks at trends in corporate IT, connecting workers to their jobs. "It turns out that the more distributed we become and the less dependent on place, the more important the quality of place becomes. It is still important to come into the office, to network and collaborate... The workplace has evolved into a club." [workplace dynamics, workplace technology, remote work, design integrity]

DeParle, Jason [2009], "Struggling to Rise in Suburbs Where Failing Means Fitting In," *New York Times,* 4/19/09: Trends of Central American Hispanic immigrants in the US – an apparent hostility toward education, 30%+ have children as teenagers. [demographics, diversity, immigration, creative talent]

Dewan, Shaila [2009], "River Basin Fight Pits Atlanta Against Neighbors," *New York Times,* 8/16/09. Without water, you have a problem. This article points toward an important water resource site, waterconserve.org, which tracks water conservation issues worldwide. [ecology, scenarios, infrastructure, real estate strategy]

Diamond, Jared [2005], *Collapse: How Societies Choose to Fail or Succeed*, New York: Viking Press [ecology, scenarios]

Dillon, Sam [2009], "Report envisions shortage of teachers as retirements escalate," *New York Times,* 4/7/09. What happens when 1,000,000 teachers retire? Most of the teachers in W.VA, VT, and ME are between 60-65. [demographics, creative talent]

Drutman, Lee and Cray, Charlie [2004], *The People's Business: Controlling Corporations and Restoring Democracy,* San Francisco: Berrett-Kokhler [organizational learning, systems thinking]

Duggan, William [2007], *Strategic Intuition: The Creative Spark in Human Achievement,* New York: Columbia University Press [innovation, creative talent, future mind]

_____ [2002], *Napoleon's Glance: The Secret of Strategy,* New York: Nation Books [strategic planning]

Dychtwald, Ken, Erickson, Tamara, and Morison, Bob [2004], "It's Time to Retire Retirement, *Harvard Business Review,* 3/04. "Companies are 'running off a cliff' by pushing out older workers as Boomers age." [demographic, generational dynamics, creative talent]

Dye, Renée, Sibony, Olivier, and Viguerie, S. Patrick [2009], "Strategic Planning: Three Tips for 2009," *The McKinsey Quarterly,* 4/09. Discusses the current use of scenarios in a tumultuous climate. Based on a survey, the authors find that strategic planning is in a crisis and recommend: 1] using a different strategic-planning process, 2] putting scenario planning in the spotlight, and 3] more frequent monitoring. [strategic planning, methods, scenarios]

DYG, Inc. [2001], *The New Workplace – Attitudes and Expectations of the New Generation at Work: Results of Qualitative Research*, 5/01. Social aspects of workplace like spontaneous interaction, the "forty-metre" rule, knowledge sharing...downside of remote work. [generational dynamics, recruitment and retention, workplace dynamics]

Dym, Barry, Sales, Michael and Millam, Elaine [2001], "Leveraging the Assets of Older Workers," *Training and Development Yearbook*, Paramus, NJ: Prentice Hall [generational dynamics, demographics]

"Efficiency and Beyond," *The Economist,* 7/16/09. "Behavioral economics argues that human beings...tend to extrapolate recent trends into the future." This school of thought brings

Influences and Resources

psychology together with economics in ways that confirm the value of the diverse group thinking. [future mind, mental models, myths and metaphors, diversity]

Eliot, T. S. [1942], *Little Gidding*, London: Faber and Faber [future mind, anticipatory leadership]

Ellenberg, Jordan [2009], "Massively Collaborative Mathematics," *New York Times Magazine*, 12/13/09. This is one of many thought provoking articles included in "The 9th Annual Year in Ideas." [new sciences, innovation, workplace technology]

Elsner, Alan [2002], "Data reflect many abandoned teens," *The Boston Globe,* 2/7/02. Family conflict is primary cause. Tens of thousands of throwaway children every year. [creative talent, generational dynamics]

"Employer on-line monitoring meets resistance," *Washington Times*, 6/27/09: What's the boundary between work and private space? [workplace dynamics, remote work]

Engardio, Peter [2006], "The Future Of Outsourcing: How it's transforming whole industries and changing the way we work," *Business Week,* 1/30/06: special section: [remote work, workplace dynamics, living systems]

Engelman, Robert [2009], "Population and Sustainability: Can We Avoid Limiting the Number of People?," *Scientific American*, 6/2009. Rapid population growth in some regions of the world, combined with increasing affluence and explosive growth in fossil fuel and natural resource consumption, is seriously endangering a broad range of natural systems that support life. "Slowing the rise in human numbers is essential for the planet--but it doesn't require population control." ecology, demographics, scenarios]

Enriquez, Juan [2001], *As the Future Catches You: How Genomics & Other Forces Are Changing Your Life, Work, Health & Wealth*, New York: Crown Business. Biotechnolgy can prolong life tremendously. [demographics, recruitment and retention, new sciences]

Evenson, Norma [1969], *Le Corbusier: The Machine and the Grand Design* [Planning and Cities Series], New York: Braziller [urban design]

Ferguson, Ian and associates [2002], Project Future Forum, *Knowledge Work,* Monograph. Assumes a set of drivers, e.g., trend to decentralization, focus on security. [demographics, recruitment and retention, future mind, scenarios]

Finkelstein, Jim and Gavin, Mary [2008], *FUSE: Igniting the Full Power of the Creative Economy; A 21st Century Primer for Boomers and Millennials in the Workplace,* BookSurge Publishing [workplace dynamics, generational dynamics, workplace technology]

Flint, Anthony [2003], "Suburban sprawl blamed for US obesity," *The Boston Globe,* 6/20/03. Suburban people drive everywhere. [sense of place, recruitment and retention]

Florida, Richard [2010], "America needs to make its bad jobs better," *Financial Times*, 7/6/10 [creative talent, recruitment and retention]

_____ [2005], *The Flight of the Creative Class: The New Global Competition for Talent,* New York: Harper Collins [creative talent, urban design, demographics, globalization, real estate strategy]

_____ [2002], *The Rise of the Creative Class: And How It Is Transforming Work, Leisure, Community and Everyday Life,* New York: Basic Books

Forrester, Jay [2009], The Loop You Can't Get Out Of, An interview with Jay Forrester, *MIT Sloan Management Review*, 1/8/09

_____ [1971], *World Dynamics,* Cambridge, MA: Wright-Allen Press [systems thinking, ecology, globalization]

Forsyth, Randall W. [2009], "Talkin' 'Bout My Generation," *Barron's*, 5/12/09. Recession drives boomers back into workforce. Boomers, who refused to grow up, now can't afford to retire. [generational dynamics]

Franklin, James C.[2006], "Employment Outlook 2006-16: An overview of BLS [Bureau of Labor Statistics] projections to 2016," *Monthly Labor Review* [creative talent, demographics]

Influences and Resources

Friedman, Thomas [2010], "A Word from the Wise," quoting Paul Otellini, CEO of Intel, *New York Times*, 3/3/10 [creative talent, globalization]

Galbraith, Kate [2009], "Efficiency Drive Could Cut Energy Usage by 23% by 2020, Study Finds," *New York Times*, 7/29/09. The Status Quo could be altered, but we are so deeply attached to the existing structures that even simple moves like replacing inefficient appliances become difficult to accomplish. [ecology, scenarios]

Gaudin, Sharon [2006], "Telecommuting Takes Flight at JetBlue," *Datamation*, 3/3/06. JetBlue's reservation system has life-sustaining qualities for their agents and saves the company money. [remote work, workplace dynamics, workplace technology]

Gearan, Anne [2001], "High court weighs Florida's age-bias case," *The Boston Globe*, 12/4/01: Can older workers sue over cutbacks that seem to hit them hardest?" [generational dynamics, demographics]

_____ [2004], "US women imprisoned in record numbers; 48% rise since 1995 tops that of males," *The Boston Globe*, 11/8/04. [gender issues]

Geirland, John [2006], "Buddha on the Brain: Meditation," *Wired*, 2/06. What is the impact of cultivating compassion? [future mind, scenarios]

Gerdes, Lindsey [2009], "Why new grads love Cisco," *Business Week*, 9/3/09. Recruits choose which department they'll join and which manager they'll work for. Of the more than 2,500 have gone through the Cisco Choice process since its inception in 2006, 98% were still with the company after two years; an astonishing retention rate. [creative talent, recruitment and retention]

Gertner, Jon [2009], "Getting Up to Speed," *New York Times Magazine*, 6/10/09. Discusses the prospects of a bullet train between SF and LA in 2 hrs 40 minutes. [infrastructure]

_____ [2009], "Why Isn't the Brain Green?" *New York Times Magazine*, 4/19/09. Excellent article on a group methodology used by the Center for Research on Environmental Decisions [CRED] at Columbia University. [ecology, workplace dynamics]

Gladwell, Malcolm [2008], *Outliers: The Story of Success*, Boston: Little Brown [recruitment and retention, diversity, innovation, creative talent]

_____ [2000], *Tipping Point: How Little Things Can Make a Big Difference*, Boston: Little Brown [elegant solutions, systems thinking, scenarios]

Goldberg, Carey [2006], "Too much of a good thing?," *The Boston Globe*, 2/6/06. Income rose steadily between 1965 and 2007, but happiness index remained basically flat. [workplace dynamics, recruitment and retention]

_____ [2005], MIT pools its brain power: united under a roof, neuroscientists seek breakthroughs, *The Boston Globe*, 11/30,05. "...the largest research building on campus." [creative talent, design integrity, workplace dynamics]

Goo, Sara Kehaulani [2006], "Building a 'Googley' Workforce: Corporate Culture Breeds Innovation," *Washington Post*, 10/21/06 [recruitment and retention, workplace dynamics]

Gootman, Elissa [2009], "More Children Take the Tests for Gifted Programs, and More Qualify", *New York Times*, 5/4/09. Information from the National Association for Gifted Children: [recruitment and retention, creative talent, generational dynamics]

Gore, Albert Jr. [2006], *An Inconvenient Truth*, New York: Emmaus, PA: Rodale Books [ecology, systems thinking, scenarios]

Groom, Brian [2009], "Business struggles to find talented recruits," *Financial Times*, 5/28/09. [creative talent, recruitment and retention, innovation]

Gunther, Marc [2006], "Queer, Inc.: How Corporate America Fell in Love with Gays and Lesbians. It's a Movement." *Fortune*, 12/11/06 [diversity, recruitment and retention]

Hall, Sarah [2009], The women who mean business: FT top 50 women in world business," *Financial Times*, 9/25/09. Female CEOs from around the globe. *None* from Latin America. 3% of the Fortune 500 are run by women. [gender issues, creative talent]

Influences and Resources

"Hannaford Platinum LEED Supermarket Most Environmentally Advanced Supermarket in the U.S.," iGreenBuild.com, 7/30/2009. Hannaford's green market in Augusta ME. Intriguing comments on the impact of natural light on the workers. [ecology, design integrity]

Harris, Gerald [2009], *The Art of Quantum Planning: Lesson fro Quantum Physics for Breakthrough Strategy, Innovation, and Leadership*, San Francisco: Berrett-Kohler [new sciences, anticipatory leadership]

Hebert, Josef [2009], "Smart grid promises dawn of high-efficiency energy use," *Boston Globe,* 6/7/09 [ecology, scenarios]

Helft, Miguel [2007], "In Fierce Competition, Google finds novel way to feed hiring machine," *New York Times,* 5/28/07. Companies like Google, Microsoft, and Yahoo find themselves going after the same candidates via cocktail parties, treasure hunts and hack days. [recruitment and retention, creative talent]

Hendricks, Tyche [2005], "Report: 112 languages spoken in diverse region," *San Francisco Chronicle,* 3/16/05: Persian is spoken by .6% of the people living in the Bay Area! Think about the recruitment opportunities and issues with that level of diversity. [creative talent, recruitment and retention, diversity]

Holland, John H. [1998], *Emergence: From Chaos To Order,* New York: Perseus [new sciences, future mind]

_____ [1995], *Hidden Order,* New York: Perseus. "This use of building blocks to generate internal models is a pervasive feature of complex adaptive systems. When the model is tacit, the process of discovering and combining building blocks usually proceeds on an evolutionary timescale; when the model is overt, the timescale may be orders of magnitude shorter." [methods, new sciences, scenarios]

Holtshouse, Dan [2009], "The future of knowledge workers, Part 1 and Part 2," *KM World Magazine,* 9/09. Strategic thinking about recruitment and retention is very thin. [creative talent, recruitment and retention]

Horgen, Turid, Joroff, Michael, Porter, William and Schön, Donald [1998], *Excellence by Design,* New York: John Wiley [design integrity, workplace dynamics, elegant solutions]

Hubbard, Barbara Marx [1998], *Conscious Evolution: awakening the power of our social potential,* Novato CA: New World Library.

Hudson, Audrey [2006], "Making Water from Thin Air," *Wired Magazine on-line: Science/Discoveries/News*

Inayatullah, Sohail [1998], "Causal layered analysis: Poststructuralism as method," *Futures*, 30:8. "Causal layered analysis is offered as a new futures research method. It utility is not in predicting the future but in creating transformative spaces for the creation of alternative futures. Causal layered analysis consists of four levels: the litany, social causes, discourse/worldview and myth/metaphor. The challenge is to conduct research that moves up and down these layers of analysis and thus is inclusive of different ways of knowing."

Ing, David [2009], "Extending the legacy of social ecology into the emerging science of social systems," *Coevolving Innovations in Business and Information Technologies [blog],* 9/8/09. Powerful overview of the scenario field, placing scenarios in the larger context of social system theory. Abstract, but largely worth the effort. [scenarios, future mind, new sciences]

Jantsch, Erich [1980], *The Self-Organizing Universe,* New York: Permagon [new sciences]

Jayson, Sharon [2009], "'Flocking' behavior lands on social networking sites," *USA Today,* 9/27/09 [systems thinking, workplace dynamics]

Johnston, Brian [2009], "A Close-up View of Several Dahlia Hybrids," and "A Close-up View of the Balloon Flower," Microscopy-UK.org, Amazing pictures of flowers based on five pointedness. [systems thinking]

Kahane, Adam [2010], *Power and Love: A Theory and Practice of Social Change,* San Francisco:

Influences and Resources

Berrett-Koehler [workplace dynamics]

Kaku, Michio [2004], *Parallel Worlds: The Science of Alternative Universes and Our Future in the Cosmos*, London: Allen Lane [new sciences, future mind, scenarios]

Karash, Richard [2005], *Systems Thinking: A Language for Team Learning and Leadership.* Unpublished. [systems thinking, anticipatory leadership, organizational learning]

Kauffman, Stuart [1991], "Antichaos and Adaptation" *Scientific American,* 8/91. Living systems have fundamental rules that are sometimes deeply counterintuitive. An organization positions itself for the unexpected by just doing what comes "naturally." [new sciences, living systems, mental models, myths and metaphors]

Keifer, Charles, Charbit, Ron and Manning, Ken [2009], Thought piece on insight and thinking based on insight at www.insightmanagementpartners.com/insightthinking.html [innovation, creative talent, future mind]

Kenner, Hugh [1973], *Bucky: A Guided Tour of Buckminster Fuller,* New York: William Morrow [design integrity, systems thinking]

"Kenyan women hit men with sex ban," *BBC News,* 4/29/09 [gender issues]

Kilborn, Peter [2009], *Next Stop, Reloville: Life Inside America's New Rootless Professional Class,* New York: Times Books. A tale about the lives of people whose working lives have led them to have no home base. There seems to be a high correlation here between organizations that are male-dominated and male oriented and a Reloville mentality. "[Many women] feel trapped. They want to go back to their extended families. They've lost their sense of uniqueness. They try to become what they think everybody else is…They create a facade for themselves." [creative talent, gender dynamics, sense of place]

Kleiner, Art [2009a], "The Thought Leader Interview: Gretchen Daily," *Strategy + Business,* Winter 09

_____ [2009b],"The Thought Leader Interview: Tim Brown," *Strategy + Business,* Autumn 09. All 550 members of the IDEO workforce "grew up making or working with beautiful things." Organizations are taking responsibility for the larger context. He cites Shimano as a firm that "doesn't even release bikes in markets unless local governments commit to safe-cycling campaigns for the initial launch" because biking and traffic safety are so integrally connected. [design integrity, elegant solutions, creative talent, ecology, urban design, systems thinking]

Kotkin, Joel, *Does Place Matter: New Concepts for an Evolving Economic Environment,* www.lajollainstitute.org/Forum/place.htm. "We are arriving at a "placeless society." Placelessness and the rise of 'edge cities [creative talent, urban design, sense of place, real estate strategy]

_____ [1997], "The Post-Mall World: At a time when cities look more and more alike, emphasizing a metropolis' unique assets and identity can revitalize its urban fabric," *Los Angeles Times,* 6/1/97. [creative talent, urban design, sense of place, real estate strategy]

Kristof, Nicholas and WuDunn, Sheryl [2009], "Liberation: A 21st Century Manifesto," *New York Times Magazine,* 8/17/09. The oppression of women worldwide is *the* human rights cause of our time. Their liberation could help solve many of the world's problems, from poverty to child mortality to terrorism. Women own only 1% of the property in the world! [gender issues, creative talent]

Kunstler, Barton [2004], *The Hothouse Effect: Intensify Creativity in Your Organization Using Secrets from History's Most innovative Communities,* New York: AMACOM [scenarios, workplace dynamics]

Lanier, Jaron [2010], *You Are Not a Gadget: A Manifesto,* New York: Alfred A. Knopf [workplace technology]

László, Ervin [2006], *Science and the Re-enchantment of the Cosmos: The Rise of the Integral Vision of Reality,* Rochester, VT: Inner Traditions [new sciences]

_____ [1996], *The Whispering Pond: A personal guide to the emerging vision of science,* Boston: Element [new sciences]

_____ [1972], *The Systems View of the World,* New York: George Brazilier [systems thinking]

Influences and Resources

Laudal, Terry [2007], "The Deeper Benefits of Going Green: More than Just Buildings," *Greenbiz,* 10/25/07. The contribution that buildings make to envirionmental considerations:
- 65% of electricity consumption
- 40% of greenhouse gas emissions
- 30% of raw materials use [i.e., 136 million tons annually]
- 50% of ozone-depleting chlorofluorocarbons
- 12% of potable water, five billion gallons, used daily
- 12% of the land use [ecology, infrastructure, real estate strategy]

Leland, John [2010], "Imagining Life without Oil, and Being Ready," *New York Times,* 6/5/10 [ecology, scenarios]

Leopold, Evelyn [2002], "UN points to aging of world's population," *The Boston Globe,* 3/28/02. People around the globe are having fewer children and living longer, turning large parts of the developing world into aging societies. [generational dynamics, recruitment and retention, creative talent]

Lesonsky, Rieva [2010], "Women Business Owners to Lead Nation in Job Creation," *Small Business Trends,* http://smallbiztrends.com/2010/01/women-business-owners-to-lead-the-nation-in-job-creation.html

Lewin, Tamar [2009], "MIT Taking Student Blogs to Nth Degree," *New York Times,* 10/2/09. MIT is "the closest you can get to living in the Internet!" Organizations employing talented young people had better get used to their technology and culture. [generational dynamics, demographics, workplace technology, creative talent]

Lewis, Diane [2004], "For younger workers, family matters," *The Boston Globe,* 10/6/04. Generations X and Y placing less emphasis on careers, study says. [generational dynamics, demographics, sense of place]

Linklater, Richard (director)[2001], *Waking Life,* Fox Searchlight Pictures. Animated film follows a young man as he floats in and out of philosophical discussions with a succession of eccentrics and passionate thinkers, uncertain whether he's conscious or dreaming. [future mind]

Lohr, Steve [2009], "For Today's Graduate, Just One Word: Statistics," *New York Times,* 8/6/09. Many more people are attending the statistics profession's annual conference than in recent years. [creative talent, recruitment and retention]

Lynch, Sara [2009], "America's Most Stressful Cities," *Forbes,* 8/20/09. Chicago comes in #1. Traffic is one of the key elements. [creative talent, real estate strategy, infrastructure]

Mandel, Michael [2009], "Innovation Interrupted," *Business Week,* 6/15/09. US Innovation has failed to realize its promise. Low technological innovation as compared to previous eras. [workplace technology, innovation]

Marien, Michael [2010], *GlobalForesightBooks*. Abstracts of current affairs books of interest to futurists and anticipatory leaders; www.GlobalForesightBooks.org. [future mind, scenarios, new sciences]

_____ [1979-2008], *Future Survey,* Washington, DC: World Future Society. 21,000 abstracts of books related to the future. *The Wilson Quarterly* described this as the preeminent source of futures information in the United States. [future mind, scenarios, new sciences]

Markoff, John [2009], "The Coming Superbrain," *New York Times,* 5/24/09. The acceleration of technological progress may be leading to "the edge of change comparable to the rise of human life on Earth." Artificial intelligence may become both self-aware and superhuman, capable of designing better robots faster than humans can today. [scenarios, new sciences]

Meadows, Donella [2000], "If the World Were a Village," *The Millennium Whole Earth Catalog*. Great stuff about the status of income and power distributions worldwide. If the world was a village of 100 people, only 1 person would have a college education; 50% of all adults would be illiterate. [demographics, creative talent]

Michaels, Ed et al. [2001], *The War for Talent,* Harvard Business School Press [immigration,

Influences and Resources

recruitment and retention, creative talent]

Millen, David [2000], "Rapid ethnography: time deepening strategies for HCI field research," *ACM Portal*. Based on the research work of Millen, Schriefer, et al., this abstract describes the process of using ethnography in a way that speeds up data gathering. [methods]

Millen, David, Schriefer, Audrey, et al. [1997], "Thinking Spaces: The Internet and Work" *AT&T/Bell Labs*, Primary research on the impact of the Internet and other technologies on the way people work and the work environment. [design integrity, workplace technology, workplace dynamics, remote work]

Milne, Richard [2009], "Crisis and climate force supply shift," *Financial Times,* 8/10/09. "US and European operations are more likely to use Mexico and Eastern Europe than China for manufacturing than previously." [scenarios, globalization]

_____ [2009], "Skirting the boards," *Financial Times,* 6/15/09: As a result of an equality law, nearly half of Norway's non-executive directors are female – a shift some say is improving scrutiny in a way that would benefit companies elsewhere. Women as percentage of boards: Scandinavia is on top, US ranks sixth. [gender issues, creative talent]

Miscovich, Peter [2007], "The New Knowledge Workplace," *McKinsey Company,* monograph. Managing intellectual assets. "[A] knowledge workplace program integrates a company's organization, technology and workplace design to support knowledge worker's effectiveness by increasing knowledge sharing, knowledge networking and innovation." [workplace technology, workplace dynamics, design integrity]

Mulligan, Casey B. [2010], When Will Women Become a Work-Force Majority?, *New York Times*, 6/23/10 [gender issues, creative talent, recruitment and retention]

Mullins, Luke [2009], "The Home of the Future: 8 Innovations in Store," 7/29/02. [remote work]

Navarro, Marco [2009], The Soul of Your Organization, *Presencing Institute Community site,* 6/13/09. A number of links to work and workplace themes. [creative talent, design integrity, workplace dynamics]

Norman, Donald A. [1999], *The Invisible Computer: Why Good Products Can Fail, the Personal Computer Is So Complex, and Information Appliances Are the Solution,* Chapter 9: Human-Centered Product Development. Cambridge, MA: MIT Press [workplace technology]

"Number of Home-Schooled Students Rises," *New York Times National Briefing,* 8/4/04. 1.1M students taught at home. [creative talent]

Obama, Barack [2010], Forum On Workplace Flexibility: Closing Session, U.S. Office of Personnel Management, 3/31/10 [remote work, creative talent]

Oshry, Barry [2010], unpublished manuscript Power & Love, Boston: Power and Systems Training

_____ [1999], *Leading Systems: Lessons from the Power Lab,* San Francisco: Berrett-Koehler [systems thinking, workplace dynamics]

_____ [1996], *Seeing Systems: Unlocking the Mysteries of Organizational Life,* San Francisco: Berrett-Koehler [systems thinking, workplace dynamics]

Oswald, Andrew [2000], "The hippies were right all along about happiness," *Financial Times,* 1/19/06: "Can't buy me love!" Rates of depression increasing in industrialized world [workplace dynamics, recruitment and retention]

Owen, Harrison [1997], *Open Space Technology: a User's Guide.* San Francisco: Berrett-Koehler [methods]

Patnaik, Dev [2009], "Forget Design Thinking and Try Hybrid Thinking" *Fast Company,* 8/25/09. Integrating design into corporate decision making. Emphasizes multidimensional thinking called "hybridity." [design integrity, future mind]

Pear, Robert [2009], "Doctor Shortage Proves Obstacle to Obama Goals," *New York Times,* 4/27/09. Primary care providers in short supply. [creative talent]

Influences and Resources

Perez, Sarah [2009], "The Technology Generation Gap at Work is Oh So Wide," *Read Write Web*, 4/24/09 [generational dynamics, workplace technology]

Petersen, John L. [1997], *Out of the Blue: Wild cards and other big future surprises: how to anticipate and respond to profound change*, Berkeley Springs, WV: Arlington Institute

Pink, Daniel [2009], Dan Pink on the surprising science of motivation, TED Presentation, 7/09. Intrinsic motivation is based on 1] autonomy – having control over our own lives, 2] mastery – being really good and getting better at something and 3] purpose – being in the service of something larger than ourselves. [workplace dynamics, creative talent]

_____ [2005], *A Whole New Mind: Moving from the Information Age to the Conceptual Age*, New York: The Berkley Publishing Group. [future mind]

Powell, John [2003], "Why Are We So Still Divided by Race?" *The Boston Globe*, 3/23/03. A recent report by the US Census Bureau, "Racial and Ethnic Residential Segregations in the United States: 1980-2000," found American housing still largely segregated, with the worst segregation in the Northeast and Midwest. [creative talent, diversity]

Preston, Julia [2009], "Mexican Data Show Migration to U.S. in Decline," *New York Times*, 5/14/09. Recession zaps immigration flows. [immigration, recruitment and retention]

Quinn, Robert E. [2004], *Building the Bridge as You Walk on It: A Guide for Leading Change*, San Francisco: Jossey-Bass [anticipatory leadership, systems thinking]

Ramo, Joshua Cooper [2009], *The Age of the Unthinkable*, New York: Little, Brown and Company [scenarios, future mind]

Ray, Paul H. [1997],"The Emerging Culture," *American Demographics*, 2/97. "Nearly one in four American adults live by a new set of values...believe in environmentalism, feminism, global issues and spiritual searching...scattered across the country and found in all social groups." [creative talent, demographics, ecology, future mind]

Ritchel, Matt [2009], "When There is No Hand Left to Hold the Wheel," *New York Times*, 10/1/09. "Real estate brokers, pharmaceutical sales people, entrepreneurs, marketers and other say they have little choice but to transform their cars into cubicles. For blue collar workers the demands to stay connected while driving are often imposed by their bosses. [remote work]

Rock, David [2009], "Managing with the Brain in Mind," *Strategy + Business,* 8/27/09. Our brains are social organs. We feel connectedness and ostracism in the same places in the brain where we feel physical pleasure and pain. Being part of a life-sustaining organization is a pleasurable, connected experience. [new sciences, living systems, workplace dynamics]

Rodriguez, Cindy [2001], "Widowers Peak As Males Live Longer, More Are, Unexpectedly, All Alone," *The Boston Globe,* 9/10/01 [demographics, creative talent]

_____ [2001], "Populace Aging, Changing its Ways," *The Boston Globe,* 5/15/01 Topics: demographics, creative talent]

_____ [2001], "Immigrants Rejuvenate Population," *The Boston Globe,* 5/15/01: Nearly 40% of US immigrants in the last decade were between 10-19 years old. [immigration, demographics, diversity, creative talent]

_____ [2000], "Sun Belt gains seen in 2000 census data," *The Boston Globe,* 12/28/00 [recruitment and retention, creative talent]

Rosenthal, Elisabeth [2010], "Our Fix-It Faith and the Oil Spill," *New York Times,* 5/30/10 [scenarios, ecology, mental models, myths and metaphors]

_____ [2009], "In German Suburb, Life Goes On Without Cars," *New York Times,* 5/11/09 [infrastructure, ecology, remote work, urban design, sense of place]

Ryzik, Melena [2009], Mapping the Cultural Buzz: How Cool Is That?, *New York Times Art & Design*, 4/6/09. An analysis of a creative class community. [creative talent]

Influences and Resources

Sales, Michael, [2008], "Leadership and the Power of Position: Understanding Structural Dynamics in Everyday Organizational Life," *Business Leadership: A Jossey-Bass Reader,* San Francisco: Jossey-Bass [workplace dynamics, systems thinking, scenarios]

_____ [2006], "Understanding the Power of Position: A Diagnostic Model," *Organization Development: A Jossey-Bass Reader,* San Francisco: Jossey-Bass [aniticipatory leadership

_____ [2006], "Futures Thinking by Middle Managers: A Neglected Necessity," *The Systems Thinker*, Volume 17, Number 19 [systems thinking, future mind, strategic planning]

_____ [2004], "How to mentor non-family employees," *Family Business Magazine Mentoring Handbook,* Philadelphia: Family Business Publishing [workplace dynamics]

_____ [1996], "Designing Magnetic Organizations in an Age of Insecurity," *The Employee Relations Bulletin*, Numbers 1856 on 5/21 and 1857 on 6/7 [living systems, design integrity, elegant solutions, workplace dynamics, creative talent]

_____ [1990], "A New Model for Organization Design and How to Use It in the Environmental Industry," *Environmental Business Journal*, May'09 [ecology, design integrity, living systems]

_____ [1984], "Action Skills for Radical Democratic Organizations: A case study of the Metropolitan Artists Union," Doctoral dissertation, Harvard Graduate School of Education. [methods, strategic inquiry]

Sales, Michael and Shuman, Sheila [2002], *Facilitation for Results,* Newton MA: New Context Consulting & Sheila Shuman Associates [methods]

Salter, Chuck [2004], "Calling JetBlue," *FastCompany,* 5/1/04. JetBlue's approach to reservations and customer service [remote work]

Sandbu, Martin [2009], "The Iraqi Who Saved Norway from Oil," *Financial Times,* 4/29/09. An Iraqi geologist helped his adopted country, Norway, cope with the discovery of oil – and made the nation's fortune. [immigration]

Savage, Anika Ellison – see also Schriefer, Audrey Ellison

Savage, Anika and Sales, Michael [2008], "The Anticipatory Leader: futurist, strategist and integrator," *Strategy & Leadership,* Volume 36, Number 6. "Extraordinarily effective leaders can see a wide range of future possibilities, know how to take advantage of emerging opportunities while avoiding threats and engage their organizations to better navigate through dilemmas and challenges." [anticipatory leadership, living systems, methods]

Scharmer, Otto C.[2007], *Theory U: Leading from the Future as It Emerges*, Cambridge MA: SOL [methods, future mind, living systems, anticipatory leadership]

_____ [2005], "Presence in Action: An Introduction to Theory U," Keynote address at the Society for Organizational Learning Global Forum, Vienna, 9/13/05. Includes a 12 step Presencing practice. [methods]

_____ [1999], "Imagination becomes an organ of perception," dialog on leadership, 7/14/99. Based on a conversation with Henri Bortoft. [new sciences, future mind, living systems, anticipatory leadership]

Schlender, Brent [2004], "Peter Drucker Sets Us Straight: jobs, debt, globalization, recession," *Fortune*, 1/04. The US is importing 2-3 times as many job as we export. [immigration, globalization, creative talent, recruitment and retention]

Schmid, Randolph E. [2005], "More Unmarried Women, But Fewer Teens, Giving Birth, "*The Boston Globe,* 10/29/05. [gender issues, workplace dynamics, creative talent]

Schön, Don [1973], *Beyond the Stable State,* New York: W.W. Norton. Brilliant analysis of the dynamic nature of interacting forces leading to a new way of learning and knowing [methods, organizational learning, anticipatory leadership]

Schriefer, Audrey Ellison [now Savage, Anika Ellison] [2005], "Workplace Strategy: What is it and why you should care," *Journal of Corporate Real Estate,* Volume 7, Number 3 [remote work, design integrity, workplace dynamics, real estate strategy]

Influences and Resources

_____ [2001], "Beyond cyberspace: the workplace of the future," *Strategy & Leadership*, Volume 29, Issue 1. How technology will affect where we live and how we work. [remote work; design integrity, workplace dynamics, real estate strategy]

_____ [1998], "Introducing Scenarios to the Corporation: Alternatron 2010 at Unicom," Chapter 23, *Learning from the Future, Competitive Foresight Scenarios*, New York: John Wiley & Sons [methods, strategic planning, scenarios, systems thinking]

Schriefer, Audrey Ellison and Sales, Michael [2006], "Creating strategic advantage with dynamic scenarios," *Strategy & Leadership,* Volume 34 Issue 3. How the Dynamic Scenario Learning Process [Structural Dynamics] works in practice. A case showing how a bank used the process to discover crucial interrelationships to manage the technology sector collapse and the 9/11 terrorist attacks. The process is both a strategic planning tool and an instrument of leadership. [methods, strategic planning, scenarios, systems thinking]

Schriefer, Audrey Ellison and Ganesh, Jyoti [2002], "Putting CRE executives in the driver's seat: Information technology tools enable new possibilities," *Journal of Corporate Real Estate, Special Issue on IT: New Tools and Strategies for CRE*, Volume 4, Number 3 [workplace technology, real estate strategy, design integrity]

Schumpeter [2010], "The Silver Revolution: Managing an Aging Workforce," *The Economist,* 2/4/10. On how to survive the silver tsunami [workplace dynamics, diversity]

_____ [2009], "Hating what you do," *The Economist,* 10/8/09. Work-driven suicide at French Telecomm [workplace dynamics, living systems]

Schwartz, Peter [2003], *Inevitable Surprises,* New York: Gotham Books [future mind, scenarios, mental models, myths and metaphors]

_____ [1991], *The Art of the Long View*, New York: Currency Doubleday [future mind, methods, strategic planning, scenarios]

Semple, Kirk [2009], "Applications for Foreign Worker Visas are Down," *New York Times,* 4/9/09. Visas continue to be available for scientists as a result of economic downturn.[demographics, immigration, creative talent]

Sciolla, Joe [2005], "High Tech's Incredible Shrinking Office," *Mass High Tech,* 2/4 & 20/05. Technology hinders real estate recovery, but also gives office tenants more leverage. [real estate strategy, remote work]

Senge, Peter [1990], *The Fifth Discipline*, New York: Currency/Doubleday [methods, organizational learning, systems thinking]

Senge, Peter, et al. [2008], *The Necessary Revolution: How Individuals and Organizations are Working Together to Create a Sustainable World,* New York: Doubleday [systems thinking, ecology]

_____ [2004], "Awakening Faith in an Alternative Future," *Reflections: The SoL Journal of Knowledge, Learning and Change*, Volume 5, Number 7 [scenarios, systems thinking, ecology]

Shellenbarger, Sue [2009], "The Next Youth-Magnet Cities," *Wall Street Journal On-Line,* 10/30/09 [creative talent, real estate strategy]

Slywotzky, Adrian [2009], "How Science Can Create Millions of Jobs," *Business Week,* 9/7/09. "Science is a crapshoot. It depends on the efforts of hundreds of people with high IQs, PhDs, deep curiosity and strong work ethic--not to mention serendipity. It also takes a certain critical mass...in infrastructure, which means lab support, equipment, and instrumentation. It takes open communication among peers and other subtle but critical cultural factors. Success requires...a culture that attracts and awards the best minds." [creative talent, methods, living systems, workplace dynamics]

"A Smarter Planet," www.ibm.com/smarterplanet. An IBM website lays out the activities of the corporation in cities around the world to make them more inhabitable and efficient. [urban design, infrastructure, real estate strategy, workplace technology, creative talent]

Smith, Diana McLain [2008], *Divide or Conquer: How Great Teams Turn Conflict into Strength,* New

Influences and Resources

York: Portfolio [workplace dynamics, organizational learning]

Stern, Stefan [2006], "Is this the end of corporate strategy?," *Financial Times*, 3/6/06. "Shareholders are running out of patience with businesses that seem to be floundering in a time of rapid change." [methods, strategic planning]

Sternberg, Esther M., M.D. [2009], *Healing Spaces: The Science of Place and Well-Being* [design integrity, sense of place, new sciences, living systems]

Stewart, George [1949], *Earth Abides,* New York: Random House

Stodghill, Roger II [2009], "The Coming Job Bottleneck: What an Aging Workforce Means for Everyone Else," *Business Week*, 3/4/09. People are aging more slowly; working later in life. [generational dynamics, creative talent, recruitment and retention]

Strogatz, Steven [2003], *Sync: The Emerging Science of Spontaneous Order,* New York: Hyperion [new sciences]

"Subcontinental drift: Westerners beefing up their resumes with a stint in India," *Business Week*, 1/16/06. A trend in Westerners at work in India for technology and outsourcing companies. [globalization, creative talent, recruitment and retention]

Surowiecki, James [2004], *The Wisdom of Crowds: Why the Many Are Smarter Than the Few and How Collective Wisdom Shapes Business, Economies, Societies and Nations*, Boston: Little, Brown [workplace dynamics, elegant solutions, anticipatory leadership]

Taleb, Nassim Nicholas [2007], *The Black Swan: The Impact of the Highly Improbable,* New York: Random House [scenarios, future mind]

Tapscott, Don [2009], *Grown Up Digital: How the net generation is changing your world,* New York: McGraw-Hill. Net Geners are innovators who:
- want freedom in everything they do, from freedom of choice to freedom of expression
- love to customize, personalize.
- scrutinize, expect and demand transparency and information
- look for corporate integrity and openness when deciding what to buy and where to work
- want entertainment and play in their work, education and social life
- are collaborative and relational
- need and demand speed
- experience high levels of anxiety when separated from technology

Labor shortage in science and engineering is acute. [generational dynamics, workplace technology, recruitment and retention, creative talent]

"US Hispanics overtake blacks," *Financial Times,* 1/22/03. [recruitment and retention, creative talent, diversity]

Uchitelle, Louis [2009], "Despite Recession, High Demand for Skilled Labor," *New York Times*, 6/23/09 [creative talent, recruitment and retention]

van Maanen, John [1988], *Tales of the Field: On Writing Ethnography,* Chicago: University of Chicago Press. [methods]

van Mee, Juriaan and Vos, Paul [2001], "Funky offices: Reflections on office design in the 'new economy," *Journal of Corporate Real Estate,* Volume 3, Issue 4 "To employees, a 'non-stop party atmosphere,' with its obligatory social events, may be almost as tyrannical as old-economy rules about wearing a tie and calling their boss 'sir...In the Netherlands, corporate culture is deeply rooted in the Protestant work ethic." [design integrity, mental models, myths and metaphors, generational dynamics]

Vascellaro, Jessica E. [2009], "Google's Schmidt on What Sets Silicon Valley Apart,"*Wall Street Journal On line*, 11/5/09. "...when you walk through Silicon Valley, the majority of the people do not look like WASP-y Americans." [creative talent, recruitment and retention, diversity]

Vinge, Vernon [1993], "The Coming Technological Singularity," Vision-21 Symposium, NASA Lewis Research Center and the Ohio Aerospace Institute, March 30-31, 1993 [scenarios, new sciences]

Influences and Resources

Vogt, Jay [2009], *Recharge Your Team: The Grounded Visioning Approach,* Westport: Praeger [methods, anticipatory leadership, living systems]

Waldrop, M. Mitchell [1992], *Complexity: The Emerging Science at the Edge of Order and Chaos,* New York: Simon & Schuster. [new sciences]

Ward, Ed and Schriefer, Audrey Ellison [1998], "Dynamic Scenarios: Systems Thinking Meets Scenario Planning," Chapter 8, *Learning from the Future, Competitive Foresight Scenarios*, New York: John Wiley & Sons [scenarios, systems thinking]

Ward-Perkins, Bryan [2009], "Call this a recession? At least it isn't the Dark Ages," *Financial Times,* 12/23/09 [scenarios, mental models, myths and metaphors]

Waters, Richard [2006], "US group implants electronic tags in workers," *Financial Times,* 2/13/06. [recruitment and retention, workplace technology]

Wheatley, Margaret J. [1999], *Leadership and the New Science: Discovering Order in a Chaotic World,* San Francisco: Berrett-Koehler [new sciences, anticipatory leadership]

Winslow, Nathan [1996], "Introduction to Self-Organized Criticality and Earthquakes," monograph, Department of Geological Sciences, University of Michigan [systems thinking, new sciences]

Wolf, Martin [2003], "Humanity on the move: the myths and realities of international migration," *Financial Times,* 7/30/03. Population flows are a global phenomenon affecting both developed and developing countries. [immigration, globalization]

_____ [2003], "People, plagues and prosperity: five trends that promise to transform the world's population within 50 years," *Financial Times,* 2/27/03. [demographics, systems thinking, scenarios]

Yardley, Jim [2009], "On India's Railways, Women Find New Peace in the Commute, *New York Times,* 9/16/09. A distressing story of how working women in India have to have segregated travel because they are so frequently harassed by men during transport to and from their jobs. [gender issues, infrastructure, creative talent]

Zuger, Abigail [2009], The Puzzle of Spaces That Soothe, *New York Times*, 6/29/09 [design integrity, elegant solutions, living systems]

Image Credits

Dedication	Tree	iStockphoto.com/ danleap
P. 9	Junkanoo	By permission of photographer David E. Knowles, Pro Photo Co. Ltd.
P. 10	Alexander	Wikimedia, File:Christopher_Alexander.jpg, Photographer: Sajjad
P. 10	Dalai Lama	Wikimedia, File:Tenzin_Gyatzo_foto_2.jpg
P. 12	Office Interior	Photo by K. Daugela
P. 13	Paris View	Wikimedia, File:View_of_Paris,_France.jpg, Author: Kennito454
P. 13	Triangles	Wikimedia, File:Sierpinsky_triangle_evolution.png, Author: Solkoll
P. 13	Triangle	Wikimedia, File:SierpinskiTriangle.PNG
P. 14	Flower	By permission of photographer Brian Johnston
P. 32	Internet	By permission of National Academy of Sciences, © 2007, U.S.A.
P. 41	Red Dot Tree	By permission of artist Elizabeth Henderson, photograher Bob Pazden
P. 45	Day and Night	M.C. Escher's "Day and Night" © The M.C. Escher Company-Holland. All rights reserved. www.mcescher.com
P. 47	Occam	Wikimedia, File:William_of_Ockham.png, Author: (Moscarlop)
P. 48	Wave	Wikimedia, File:pdphoto.org/PictureDetail.php?mat=pdef&pg=5567
P. 53	Presidencia	Wikimedia, File:Cristina_Fern.jpg, Source: Presidencia de la N.Argentina
P. 56	Sandpile	Adapted by Anika Savage
P. 58	Frog	Adapted by Anika Savage
P. 64	Obama	By permission of the Economist
P. 65	Lanier	Wikimedia, File:Jaron_lanier.JPG, Source/Photographer: Allan J. Cronin
P. 66	Core Collapse	Wikimedia, Source: US Government. A supercomputing visualization of a core-collapse supernova.
P. 76	Brando	Wikimedia, File:Marlon_brando_waterfront_1.jpg Source: Trailer screen shot from "On the Waterfront"
P. 77	Charles	Wikimedia, File:Ray_Charles_28cropped29.jpg, photographer: Alan Light
P. 80	Happy Man	iStockphoto.com/Andrey Shadrin
P. 99	Ocean Vista	Photo by Michael Sales
P. 104	Villages	By permission of SCAN
P. 106	Chart	By permission of SCAN
P. 110	Ozymandias	iStockphoto.com/ BMPix
P. 126	Golf Cart	iStockphoto.com/ Adisai Chaturapitr
P. 126	Fork Lift	Wikimedia, File:Traigo front.jpg Source: Polska
P. 126	Times Square	Photo by Anika Savage
P. 128	Electric Car	Wikimedia, File:Buddy05.jpg, Source/Photographer: Bjoertvedt
P. 140	Jitney	Photo by Anika Savage
P. 163	Bumper	Photo by Michael Sales

Index

Alexander, Christopher   i, 10, 14, 46
AT&T   108
Anan, Kofi   2
Anticipatory Leadership
   Definition   17, 108-110
   Development through process   110, 170
   Futurist   36-38, 70, 109, 136
     Self-assessment   38
   Integrator   20-24, 92-93, 109-110, 128, 164
     Self-assessment   25
   Strategist   74-76, 93, 109-110, 153
     Self-assessment   76
Apple Computer   120
Appreciative Inquiry   100, 117, 166, 183
Archetype, defined   59
Arthur, W. Brian   74, 183
Bak, Per   54
Bortoft, Henri   97
Brando, Marlon   74
Cahill, Thomas   68
Certainties   42, 139
Change
   Approaches to   4-5, 15, 66-68
   Living systems and   2, 14, 15, 54, 183
   Pace of   6, 36, 54-56, 140
   Responses: hope or fear   57, 142
   Scenario Game Board and   65-66, 142-148
   Strategic planning and   96-97
   Structure and leverage points for   52-53
   Thinking the unthinkable   79
   Timing   5-6, 69
Charbit, Ron   78
Charles, Ray   75
Coleman, Ornette   109
Coles, Diane   102-105
Conniff, Richard   93
Convening   15, 19-33, 128-135
   Articulating the Decision Issue   31-32, 134
   Integrators and   22-27
   Preparation for   25-29, 128-130
Cooperrider, David   27, 100, 166, 183
Corporations
   Criticism of   182
   As an agent of world benefit   2, 183
Creative people and talent
   Attracting and holding   6-7, 15, 21, 76, 109, 114, 157, 173
Critical uncertainty   43, 65, 141
Daily, Gretchen   183
Dalai Lama   10, 47
Data analytics   118-119
Dator, Jim   59-60
Deal, Terrence   115

Decision Issue
   Articulating   31-32, 134
   Critical uncertainties and   43, 65, 83
   Example   81
   Framing questions and   39, 41, 137
   Scanning and   119
   Time horizon   32, 134
Defensive routines   93
De Geus, Arie   3, 108
Design integrity   11-12, 23-24, 175-176
Diamond, Jared   64
Discovering   18, 73-90, 151-163
   Analyzing future possibilities   77-79
   Articulating scenarios and   80-82, 153-158
   Developing and testing strategies   86-89, 159-162
     Strategy matrix   88, 161
   Strategists and   74-77, 152
Duggan, William   75, 114
Dylan, Bob   113
Elegant solutions   12-14, 24, 36, 53, 109, 176
Embodying   18, 91-106, 164-169
   Embedding and   95-97
   Integrators and   92-94, 164
   Whole system events   98-106, 165-169
Executive briefings   71, 90, 106, 150-151, 163, 169
Evolution of work   181-184
Exploring   35-71, 135-151
   Fathoming structure and   51-53, 141
   Futurist and   36-38, 136
   Recognizing patterns and   46-50
   Scanning events and   39-42, 136-138
   Scenario Game Board, use of in   54-68, 142-148
Facebook   116, 120
Finlayson, Roosevelt   9
Flash transcription software   167
Forrester, Jay   52
Free Jazz   109
Fuller, R. Buckminster   3, 46, 175
Future mind   185
Gladwell, Malcolm   55, 70, 75
Google   14, 118, 120
Harman, Willis   183
Haussmann, Georges-Eugene   12, 181
Horgen, Joroff, Porter and Schon   119
Jantsch, Eric   3, 10
Kahane, Adam   110-111
Kaku, Michio   3
Keifer, Charles   78
Kunstler, Barton   94
Le Corbusier   181-182

Index

Lamola, Angelo 112
Lanier, Jaron. 63
Laszlo, Ervin 14, 59
Leaders,
    Categories of 20
    Line leaders and work environment 21-23
Leadership in Energy and
    Environmental Design [LEED] 175
Life-Sustaining Organizations
    Anticipatory leadership and 17
    Becoming life-sustaining 15-16
    Corporations and 182-184
    Defined 6-15, 131-133
    Magnetism of 5, 7, 94, 121
    Monitoring and 114-120
    Operational plans for units to become
        97, 168-169, 172
    Organizational longevity and 108
    Personal experience of 15
    Planetary consequences
        180, 184-185
    Power and love and 110-12
    Success to the successful and 120-121
Living systems 10, 31, 50, 52, 96-97
    Defined 2-3
Marien, Michael 119
Mastery and practice 17, 75, 114, 181
Matrix, The [movie] 50
Mead, Margaret 6, 183
Method acting 74
Michelangelo 181
Myth, Metaphor and Mental Models
    As filters 69-70
Obama, Barack 9
O'Brien, Bill 2
Occam's razor 45
Organizational learning 17, 30, 92, 120
Organizational magnetism 5, 7, 94, 121
    Diagrammed 7, 21, 38, 76, 94, 112
Oshry, Barry 10, 21-23, 93, 110-112
Otellini, Paul 6
Patterns 8, 13, 14, 39, 46-50, 110, 141
Playbook
    Defined 29-30
    Coordinator 26, 32, 130, 133
    Criteria 32, 134
    Use of 71, 90, 100-101, 106, 133,
        134, 135, 138, 142-143, 149,
        158   167, 169, 171, 172-173
Power and Love 110-112
Power Lab iv
Power law 54, 120
Productivity measures 116
Remote work 12, 114, 146, 149, 176

    As a pattern 48-49
Results orientation 14-15, 176-177
Roosevelt, Eleanor 186
Sales, Michael iii-v, 10, 26, 118
Satisfaction surveys 117
Savage, Anika i-iii, 11, 12
SCAN Health Care 102-105
Scanning
    Bias in 41
    Critical uncertainties 43, 80-82,
        84-85, 139-140
    Examples 40, 136-138
    Events 39-42
    Framing questions 39, 137
    Safe bets 42, 139
    STEEEPA 40, 119-120, 136-137
Scenarios
    Archetypal 59-64
        Collapse 64
        Discipline 62
        New Reality 63
        Status quo 61
    Articulating 80-82, 153-155
    Naming 83, 156
    Narratives 83-85, 156-158
    Uses of (Examples) 77
    Living in the future 82-85, 155
    Pace of change and 54-56
    Response to change and 57-59, 142
    Scenario game board 54, 65-68,
        113, 142-148
SecondLife 99
Scharmer, Otto 53
Schriefer, Audrey 11
Self-organized criticality 54-56
Senge, Peter 20, 113, 116, 183
Shuman, Sheila 26
Sierpinski triangle 13
Signposts, indicators, and warnings 114
SpectorSoft 116
Strategic conversations 99-101, 111, 118,
    165-168
Strategic insight 42, 78
    Thinking the unthinkable 79
Strategic intuition 114
Strategic planning
    Fragmentation of traditional methods 94-95
    Organization-wide activities 98-106
    Scenarios in 86-88, 159-161
    Whole systems thinking and 96-98
Strategies
    Contingent 89, 113, 162
    Existing 89, 162
    Matrix 88, 90, 161-162

Index

Robust   88, 101, 162
Stress testing   86-87, 159-161

Strategy team
- Assembling   26-29, 129-130
- Charging   131-133
- Diversity   27-29, 183
- Facilitating   26, 32, 41-42, 128-129, 133, 165
- Ground rules   32, 134
- Responses of executives to   71, 90, 151-152, 163
- Roles   26
- Work environment   30, 131

Structural Dynamics
- Complexity and   180
- Creating a model   82-83, 156-158
- Definition   16, 77
- Anticipatory leadership and   17, 170
- Driving forces   36
- Executive tensions and   93-94
- Heart of   54-60, 142-144
- Learning and   92, 114, 120, 154
- Method acting and   74
- Open dialog and   93-94
- Roles in process   26
- Summary of process   187-190
- Theory U and   53

Structure
- The power of seeing   52-53

Success to the successful
- system archetypes   113, 120

Supply of skilled workers   7, 141
- As example   144-148

Sustaining results   16, 107-121, 170-177
- Methods for monitoring   115-120, 170-171, 173-176
- Balancing power and love and   110-112

Taleb, Nassim Nicholas   79, 175
The Temptations   180
Theory U   53
Total recall technology   118
Van Maanen, John   115
Variables   44-45, 140-141
- Patterns and   46-50

Waking Life [movie]   63
Whole Systems Thinking   8-11, 21-23, 38, 96-98, 173-175

Women's Rights
- And patterns   49-50
- As a variable   44-45
- Structure and   51-52

Made in the USA
Lexington, KY
14 January 2011